BUSINESS

SANTA FE
33

Santa Fe Railway Passenger Car Reference Series - Volume Four
Business & Special Purpose Cars

Graphic design by John R. & Jonathan Signor
Cover and frontispiece: Car 33 on the rear of train 23, at Seligman, Arizona, December 22, 1968, by Joe McMillan

The Santa Fe Railway Historical & Modeling Society Inc.
1205 S. Air Depot #101 Midwest City, OK 73110-4807

Printed in the United States of America by AgPress, Manhattan, Kansas.

ISBN 1-933587-11-3

Santa Fe Railway Passenger Car Reference Ser

Business & Special Purp

Based on the original work by Frank M. Ellington & Jose

Published by the Santa Fe Railway Historical & M

Santa Fe Railway Passenger Car Reference Series – Volume Four

Business & Special Purpose Cars

Based on the original work by Frank M. Ellington & Joseph W. Shine

Published by the Santa Fe Railway Historical & Modeling Society

Santa Fe Railway Passenger Car Reference Series - Volume Four
Business & Special Purpose Cars
Graphic design by John R. & Jonathan Signor
Cover and frontispiece: Car 33 on the rear of train 23, at Seligman, Arizona, December 22, 1968, by Joe McMillan

The Santa Fe Railway Historical & Modeling Society Inc.
1205 S. Air Depot #101 Midwest City, OK 73110-4807

Printed in the United States of America by AgPress, Manhattan, Kansas.

ISBN 1-933587-11-3

Foreword

It has been more than thirty years since this volume was originally published. Since that time many of the contributors—and indeed the Santa Fe itself—have passed on. Released in limited quantities and much sought after by collectors, *Business and Special Purpose Cars of the Santa Fe* was considered important enough by the Board of the Santa Fe Railway Historical & Modeling Society to reprint. Frank Ellington and Joe Shine were again contacted and with their support it was determined to revise and update the original.

Over the years since its release in 1975, the cars continued to move through history, considerable new material has come to light, and quality printing—especially of color—is now within the reach of the Society. Thus the volume you hold in your hands is essentially an entirely new book, based on the original but considerably expanded. As the revision moved forward it was decided, in keeping with the focus on rail cars, to exclude the segments on the motorized inspection cars and Santa Fe Airways, which appeared in the original volume, and publish this material later in another form.

In 1975 the authors expressed their gratitude to the following individuals and/or organizations that had contributed images or other material from their collections. First of all, they thanked the Santa Fe people: William Burk, Vice President-Public Relations, Chicago, William B. Cox, Gene Flohrschutz and John Ellsworth, Public Relations, Los Angeles, Jerry Curto, Manager, Audio-Visual Department, Chicago, and Frank Meitz, System Chief Photographer, Los Angeles.

Public repositories which were of much assistance included the DeGolyer Library, Southern Methodist University, Dallas, Texas; the Huntington Library; San Marino, California; Iowa State University Library; Ames; Kansas State Historical Society; Topeka; the Museum of Transport; St. Louis, Missouri; the New York City Library; the Orange Empire Trolley Museum; Perris, California; the Chicago Public Library; The Smithsonian Institution; Washington D.C.; and the Spencer Research Library; University of Kansas, Lawrence.

Individuals who had contributed photos and/or information to the original work included Gene Anderson, Cecil Balentine, Blaine Bowman, Lee Burglund, Ronald Dloughy, Cyril Durrenberger, Vernon Glover, Henry Graham, Richard Hendrickson, Ray Hilner, Frank O. Kelley, Bill Kingston, Stan Kistler, Dennis J. Kogan, Ron Machalek, Edward Mahoney, Loren R. Martens, William Messecar, Joe McMillan, R. P. Middlebrook, John Moore, Larry Occhiello, Steve Patterson, Alton Parsons, John Platt, Jr., Jerry Roundy, Richard Scholz, Ralph R. Shankland, Charles Slater, Charles Trappani, Jr., Robert Wayner, Chard Walker, Will Whittaker and Harold Williams.

This revised and expanded edition of *Business and Special Purpose Cars of the Santa Fe* is the result of a collaborative effort between John Signor and Keith Jordan.

Assisting us in this effort with new photos, text and roster information, advise, and commentary were Wally Abbey, Howard Ameling, Ed Birch, Harry Briscoe, William R. Carter, Bill Childers, Frank Ellington, Kevin Eudaly, Mike Flick, Lewis Harris, Thomas G. Hoback, Phil Kauke, K.B. King, Bob Knoll, Dick Kuelbs, Jim Lancaster, Elrond Lawrence, Paula Lester, Mike Martin, Louis A. Marre, John B. McCall, Joe McMillan, Bob Melbo, Orville Pratt, Steve Patterson, Robert Pounds, Kenny Rhoads, Steve Sandifer, Warren Scholl, Bill Sheehan, Seeth "Corky" Thompson, Harold K. Vollrath, Mark R. Wayman, Charles E. Winters and Jack Whitmeyer.

The late Connie Menninger, William Martin and Nancy Sherbert of the Kansas State Historical Society and Craig Ordner of the Railroad & Heritage Museum at Temple, Texas, were especially helpful.

Jonathan Signor and Matt Zebrowski provided scans, prepared raw text for editing and otherwise helped in the editorial process. Long time Santa Fe authorities Art Riordan, Frank Ellington and John B. McCall proofed all or portions of the final product.

The old and the new: Shadow-lined Car 22, and Car 53, Mountainair, lay over on adjacent tracks at Albuquerque, New Mexico, on April 22, 1967. *-Steve Patterson*

Table of Contents

While train No. 16, the *Texas Chief*, makes a station stop at Fort Worth in December 1970, an official, presumably the Eastern Lines General Manager, takes a bit of fresh air on the platform of heavyweight Business Car 34. Meanwhile a carman, flagman and porter attend to their duties under the anxious and watchful eyes of the local trainmaster. *-John B. McCall*

1

Introduction

The railroad business car is a unique form of transportation. While there are private cars, boats and planes, very few are intended as self-sufficient office, entertaining and sleeping quarters, designed for long-distance traveling. Indeed, the closest parallel today might be the President's Air Force One. For most of the twentieth century, however, middle management and executives of the nation's railroads had at their disposal a fleet of railroad cars designed for business.

Santa Fe's fleet of Business cars started work in 1893 with two cars delivered from the Pullman Company (see chapter two). The purpose of the cars was to allow an official to travel the line, important in the day when roads barely existed as dirt paths outside of cities and towns. By having a self-contained car, an official could visit his territory at leisure, stopping and talking with employees everywhere. Officials could keep in touch with headquarters via stations along the way and, later, via telephone or radio. As such, business cars wielded the clout of the office and the mystery that surrounds it, just as any four-walled executive room in a stationary building would.

Depending on to whom one talked, life aboard a business car could be either one of luxury or misery. If you were the official, there was a chef to cook your meals and a porter to look after your personal belongs, press your clothes, change your linen (in the case of the smaller 400 series cars, a single cook/porter was assigned) and generally keep things straight. A secretary (in all cases, male) handled correspondence, mail and the usual duties of that multi-faceted title, plus some. Obviously, if you were the cook, porter or secretary, life aboard the car was different, sometimes difficult and more than a few times, probably miserable. But those that did work and serve aboard business cars all felt they were a lucky few to have experienced it.

While the cars themselves were designed with a certain amount of comfort in mind, they were not, by any standard opulent or extravagant. Utilitarian is probably a better description, keeping in mind that the definition of that word has changed in the last one hundred years. Business cars of the early twentieth century had hand-finished wood interiors, plush upholstery, porcelain and brass fixtures, all prized today when most decorative items are manufactured with the lowest costs involved. The cars assigned to top executives and officers were well appointed, since they could be used to entertain other executives, customers and celebrities.

An examination of the photos and material covering the car numbered 17 (beginning on page 57, chapter three), built in 1907 for E. P. Ripley, then President of the Santa Fe, will certainly give the reader an idea of the beauty of design, both inside and out, of these cars—yet even here, spaciousness usually connected in the mind's eye with sumptuousness, is mostly lacking. Due to clearance restrictions of track along the right-of-way, all cars were of necessity, limited to a width in the ten foot area, give or take a few inches. Examine any of the floor plans accompanying the photos to appreciate the maximum employment of the limited space available and further to bear in mind much additional storage space was utilized under the car body, as well as in the overhead interior in the clerestory section.

The car type designed for service as outlined above, such as number 17, represents the pinnacle of Business car design and appointment on the Santa Fe, while the cars built for Division Superintendents were abbreviated in length and quite different in interior appointments. The term "utilitarian" might be more applicable to these business cars but the very early wood versions (chapter three, page 71) can best be described in most cases as merely "junior editions" of their early big brothers. The more prosaic service of a Division Superintendent, where an individual occupant, home-away-from-home might be required on Company business, saw that individual generally ensconced in a car assigned to him which was, in interior appointments, every bit as well fitted out as the interiors of the longer General Officer-assigned equipment. Fine interior mahogany woods,

Plains Division superintendent J.B. Briscoe on the steps of car 409 at Amarillo, 1938. *-Harry Briscoe collection*

suitably finished and polished to a mirror-like sheen, were the order of the day. But even here, that same excellence could be found in all passenger carrying equipment in first class service, whether it be coach, diner, sleeping car, or other type. That was just the way things were done in the early days, each car reflecting high pride in accomplishment on the part of the shop craftsmen whose employment was in the coach construction shops such as in Topeka, Kansas.

With the advent of steel construction, however, and the accompanying (though not necessarily parallel) development of higher labor costs, methods and materials for interior and exterior finishes were resorted to which were much less ostentatious. The final group of Division Superintendent's Business cars (chapter five), built circa 1924, can well be termed extremely functional in design, spartan in finish, and practical to the "nth" degree. Steel plate was used throughout (including interior paneling) and an interior finish of cream color enamel was the norm. Along about the same time, steel cars were becoming the vogue for the general officer equipment and the older, wood side cars were receiving steel plating. Newer construction, though using steel plating for partitions, saw the interiors finished in a manner to simulate wood grain finish, maintaining the earlier look using more modern materials. This was due, of course, to ease of maintenance and reduced labor costs.

Business cars, if nothing else, are a reflection of the railroad and its times. Reflecting the move to lightweight, streamlined equipment for revenue service a decade earlier, the Santa Fe began purchasing similar equipment for its business fleet in 1949. The interior appointment reflected the mid-century "modern" design practices which included enameled-steel partitions, wood cabinets and tables, with simply designed furniture. The cars were essentially maintained in this style until the survivors' interiors were totally redecorated in 1990.

Life On the Cars

Ed Mahoney retired as Chief Clerk on the Santa Fe. In his early years as a secretary to superintendents and other officials on the Western Lines, Ed racked up many miles in his three years aboard Business cars attending to, well, railroad business. He felt there was "relatively little purely pleasure traveling in these cars." If an official was out in the car, even for personal use (which was allowed), some business was conducted. Ed accompanied his boss to the scenes of derailments, collisions, washouts or new construction somewhere out "on the line." Ditto to terminals where business meetings were held with other officials or a formal investigation occurred. The business car then became the office for such meetings and Ed, as secretary, wound up being the stenographer taking minutes. Ed recorded his life in a series of typed correspondence, relating life on a business car and a by-gone way of conducting work.

"My first-hand experience with these cars covers a three-year period Spring 1919 to Spring 1922. I worked as a secretary on cars 421, 19 and 4. The first two were assigned to division superintendents, the 4-spot to a general manager. In addition, I had opportunity to be in numerous other Business cars.

"In 1919 there were 46 Business cars on the roster. They varied greatly in construction and appointments but the general configuration was about the same except that some of the older superintendents cars combined the dining room and observation room. Car 98 which was assigned to the superintendent at Prescott, Arizona (and which I never saw) was only 30 feet over the end sills (meaning exclusive of the platforms). The floor plan indicates that this car had a cupola. A few of the other old ones were not much longer. Car 405 assigned to the superintendent at Slaton, Texas, and Car 409, assigned to the superintendent at Pueblo, Colorado, were only 34 feet over end sills. These cars were all wooden and had not even been equipped with electric lights-oil lamps hung from the ceilings.

"At different times, I was secretary to two of what must have been the hardest working superintendents on the Santa Fe—one at Wellington, Kansas, and the other at Las Vegas, New Mexico. They each handled a heavy volume of correspondence. Many superintendents didn't concern themselves too much with office work but these two did. It paid off, as both were later promoted while many superintendents never got any farther. Each of these superintendents had a sort of "hideaway" where we would go for two or three days to catch up with office work and where we could be relatively free from interruptions. My boss in Wellington used to go out to Belvidere, Kansas—an intersecting point of two track lines. Other than the depot and a company house or two, that was about all there was at that point. My boss in Las Vegas used to go to Lamy, New Mexico, every now and then. This was the junction point for the 18-mile branch to Santa Fe. At this point there was the station, a small Fred Harvey eating house and hotel (eight rooms, one of which the manager used), a few company houses and three or four others, including a small store. Of course we kept in touch with Division Headquarters by telegraph and, at Lamy by the dispatchers' telephone and while there received and dispatched mail.

"When railroads ran frequent local freight trains, those which stopped at every station to set out or pick up cars or to unload and load less-carload freight, one of the best opportunities for a superintendent to supervise stations along the line and to observe track conditions, etc. was to put the Business car behind the caboose on such trains. There was time at every stop for the superintendent to talk to the agent and give the depot and surroundings a check.

"At one time the Business car was used as the pay car for the district between Raton, New Mexico, and La Junta, Colorado. The pay was not disbursed in money, as in the days when pay cars were actually used, but by the regular employees' pay checks. A representative from the paymaster's office accompanied us. We used a local freight train and stopped at every station or intermediate point where a section gang was working. The men would file in, give their name, sign for the check and then have it handed them. The purpose of this one-time deal was to see if there were any "straw men" carried on the rolls, that is, non-existent employees. I don't recall that there were any irregularities.

"I recall one occasion when I was on a special train with my boss, the general manager, in Car 4. En route from La Junta, Colorado, to Denver the following cars trailed the engine (one of the 1200 class Pacifics) in this order: Car 409, Car 22 (assigned to the assistant general manager, La Junta) Car 4 and an all-steel New Haven Railroad business car occupied by Thomas DeWitt Cuyler, who was chairman of the Association of Railway Executives and a director of the Pennsylvania Railroad. Cars 4 and 22 had 6 inch I-beam underframes and steel platforms. Our distinguished visitor in the New Haven car liked to ride fast and he was taking a rather comprehensive tour of the Santa Fe. We were running just about as fast as conditions would permit. The cook in Car 409 was so terrified he wouldn't ride in it and was back in one of the other cars. I couldn't say that I blamed him. In any serious accident, that car would have collapsed like a match box. The conductor and brakeman on our special were riding in the little dining-observation room and they didn't look too happy. The oil lamps were swinging back and forth. I walked up to the kitchen end of the car and looked out on the platform. The tank of that 1200 was rocking from side to side and every once in a while a big lump of coal would come flying back.

"Santa Fe has several sections of heavy grades which, in the days of steam, required helper engines. With the exception of Raton Pass, one helper was sufficient but on Raton Pass two helpers were needed for the usual passenger train. It was the practice then to put one helper ahead of the road engine and the other on the rear, pushing. Even on Glorietta Pass, where a helper was required eastbound, a pusher was used instead of double-heading except that when a train had a business car on the rear the helper engine went on the point. One time we were getting ready to leave Lamy, on No. 10 going up to Glorietta. Because some freight train had recently run through a switch, breaking the switch points, the helper engine couldn't run around the train to get up to the head end. My boss, the superintendent, told the crew on the helper to go ahead and push on his car but to take it a little easy and just keep the slack bunched up. We had Car 19 with the steel I-beam underframe and steel platforms so I don't think any hazard was involved. On Raton Pass some special moves were necessary when handling Business cars. If we were going through to La Junta or beyond eastbound and were on No. 8 which took breakfast at Trinidad, or No. 10, which took dinner there, the Business car was placed behind the pusher. After going through the summit tunnel, the pusher with the Business car would follow the train and there would be ample opportunity to switch the car to the rear of the train during the meal stop. Westbound if we were cutting out at Raton, the practice was the same, i.e., the car would be behind the pusher which would follow the train down to Raton. At any other time, however, when we arrived at Trinidad westbound or Raton eastbound, a third helper engine would be used and this engine with the Business car would run ahead of the train as it could make better time than the heavy passenger train which followed.

For many years, Topeka Passenger Car Shops was headquarters for upkeep of the Business cars. It was from Topeka that Santa Fe selected men serve on the Business cars. This 1939 photo shows the class of the men who manned the cars. Left to right: Front Row, Lester G. Martin, William D. Cooper, Earnest Gordon, Adolph Hubbard, Charles Slaughter, Otto Bennett, George W. Cowan, John H. Lawson, and Charles Douglas; Middle Row, William A. Gilbert, Orlando W. Atkinson, Thomas E. Merrill, Tutson B. Blount, Charles French, Walter Jordan, Peter E. Roberts and Robert Perkins; Back Row, Samuel Jenkins, Carl Finney, William Gaines, Taylor Reed, George Lee, John M. Warner, William E. Allen, Frank Howard and Albertus Pogue. *-Harry Briscoe collection*

"Car 19 had a fine looking brass railing on the observation platform. When the car was returned to Las Vegas, New Mexico, following a refitting in Topeka, my boss discovered that someone had painted this railing black. This made him very unhappy. He had the car switched down to the roundhouse and had the Mechanical Dept. burn off all this black paint, restoring the natural brass appearance. Most of the cars had a mahogany finish on the woodwork but I recall that Car 22 had an oak finish. In Cars 19 and 421 my typewriter desk was in the dining room on the right side, forward. One of the two staterooms was assigned to me. They had transverse beds with a upper above. Most of the time I had the bed; occasionally someone else would share the room with me and use the upper. One reason I ordinarily kept the lower was because, just like in a home, the sheets and pillow cases weren't changed every morning. On a couple of occasions I got outranked and my boss took my lower and I the upper. On these occasions he had members of his own family in his room.

"In Car 4 my room also served as my office and all had was an upper running the length of the car. I would have preferred a transverse upper as in the secretary's room in Cars 15 and 18. The general manager didn't handle much correspondence but two superintendents which I worked for did. After a meal, with the dishes removed, the cook/porter

While I did not regularly use car 19 as a bedroom when we were in Las Vegas, I did occasionally sleep there, during the winter and on nights when I didn't have a date. On such occasions, I'd spend some time around the station watching the evening passenger trains and visiting the dispatchers' office, where a night chief dispatcher and two trick dispatchers were on duty. Then, before turning in, I'd go over to the Fred Harvey lunch room and get a bite to eat. As the 19 was parked just about 50 feet from the office, on steam heat, it seemed handier just to go to my room onboard there, rather than walk several blocks to my rooming house.

One night after I had bedded down in the 19, the old man came bursting in shouting, "Who you got in here? Who you got in here?" He actually seemed disappointed to find me alone! I had taken two or three of my lady friends into the car on occasion, just to show it off, but had never suggested we make a night of it.

I did have a couple of girls to dinner one evening in Raton. New Mexico Division superintendent F.L. Meyers had an odd habit of occasionally leaving me and the car at some point while he tore off to the other end of the division. The "grapevine" would undoubtedly report where 19 was stationed and presumably the "old man" would be in it, whereupon they would much surprised to find him showing up at some other point! It could have been that he did this to confuse the men out on the line, but he was just impulsive and did things on the spur of the moment.

Anyhow, one time when we were in Raton, the boss announced he was going to make a quick trip to Albuquerque or some place and I saw him leave, so I knew I'd be safe for a day at least.

I knew a young lady there and thinking it would make an impression on her, invited her to have dinner with me on the car that evening. She was a little hesitant about accepting, but finally agreed on the condition that she could bring her sister. I wasn't too enthusiastic about this but it was better than nothing, so they both came. We had a big steak dinner with all the trimmings, compliments of the Santa Fe. This meal set me back all of fifty cents—a tip for the cook. *-Edward Mahoney*

would put a big spread on the table about like a large throw rug today and we would put our files and correspondence on it. At the front of the dining room was the sideboard or buffet. The boss usually sat just in front of that when we were eating, riding backward. I sat at the foot of the table riding forward and guests, if any, sat at the sides. I suppose the reason the boss sat at the head of the table was so he could be served first, from the kitchen.

"We drew all of our supplies from the five Harvey Houses on the division: La Junta, Trinidad, Las Vegas, Lamy and Albuquerque. The cook would sign for what he wanted and once a month we'd get a bill from Fred Harvey. Into this, I would divide the number of meals that month, thereby arriving at an average cost per meal. We reported this through channels to the General Manager and each superintendent would strive to make a good showing. As the cost didn't go up in proportion to the number of meals, the more meals served, the better the showing. Total cost apparently wasn't involved.

"Not many cars had bathing facilities at that time. If there were any which did I never saw them. For one thing, back in those days, most people thought that a bath once a week was sufficient. I can recall after having been out on the road for several days, that when opportunity presented, I would go up to the nearest barber shop. Practically all of them had a bathtub in a rear room where customers could bathe for a charge of 25 cents.

"Looking back at those days I marvel now at how completely self-contained these Business cars were. At a terminal, during chilly or cold weather, the cars would be placed on a steam heat line but this was the only outside facility used. There were no electrical connections or battery rechargers. When we were side-tracked at some point where there was no steam line or when we were riding on a freight train and the weather required it, heat was provided by the efficient Baker heater at the front of the car. I never saw coal being delivered for this heater or the batteries being recharged or replaced or the water tanks filled. It was just something I took for granted. At that time the practice of hanging the "honey pots" immediately under the toilets beneath the car had not developed. For important calls of nature when we were at some station or terminal, we used the station or office building facilities.

"Periodically business cars would go into Topeka Shops for a general overhaul and repainting. I don't know whether general managers got a relief car during that period—I wasn't with the general manager in Amarillo long enough to run into this situation, but I do know that superintendents and assistant general managers just did without during this period which, it seems to me, took about, a month. It was a happy day for me when the car returned, because I liked to travel. The car would come back almost new with shiny varnish both inside and out, worn upholstery repaired or replaced, etc. The car would have a sort of 'new' smell, not similar to, but somewhat like, that smell of a new automobile.

"Several officials in Chicago, in the upper echelon, in other than the Operating Department, had the privilege of using a Business car on their occasional trips. Use of a car is not necessary for inspection purposes, etc. as in the case of the Operating Department, but it is certainly a convenience and adds enormously to the prestige of this official when he is in other cities and having meetings and conferences with executives in private industry, civic leaders or government officials. Traffic Department officials in particular use the car for entertainment purposes. Many years after I had left the Santa Fe and had a responsible position with private industry and later with the U.S. Government I was treated to meals not only in Santa Fe Business cars but in those of the Burlington, Missouri Pacific and Rio Grande. I can safely say that this makes a much bigger impression on a shipper than being invited to lunch at a fine exclusive club."

Mahoney also talks about the cars being used for purposes other than strictly business. Typical pursuits would be fishing outings, getting away from it all or even entertaining a date or two (see sidebar).

"My boss at Las Vegas liked to go trout fishing and occasionally we would set out at Rowe, New Mexico, where we had a hike of about a mile down to the Pecos River. His preference, however, was for Ute Park, about sixty miles west of Raton, at the end of a branch line. This is beautiful mountainous country and in the 1920s there was little other than the depot and the agent's house. This agent, in addition to his railroad duties, was also the local game warden, the postmaster and the proprietor of a small general store! About the only outbound traffic was a car of logs or lumber and occasionally a gondola from some mine back in the mountains. We would go out there on the mixed train from Raton, arriving about 1:00 p.m. When going up to Ute Park, I always tried to arrange my work so I didn't have anything to do for the last eleven and half miles from Cimarron. I liked to sit out on the rear platform with my feet propped up on the brass railing, watching the scenery as we ascended Cimarron Canyon on a two percent grade, where we crossed and re-crossed the river on many wood trestles, listening to the squeal of the flanges on the sharp curves and smell the heady aroma of pine, sage and good, soft coal smoke. The train returned to Raton the same afternoon but we would occasionally layover for 24 hours. The rushing Cimarron River was only a stone's throw from the track on which the Business car was spotted. Other stops and business were always connected with these brief fishing trips.

"When I was secretary to the general manager at Amarillo, Texas, I can recall only one trip which might have been considered for personal reasons. In recognition of his wedding anniversary, we went from Amarillo to Colorado Springs, Colorado, for a few days. Naturally his wife went along, but this was not unusual. Sometimes a wife would accompany her husband even if the trip was all business. On the occasion of this Colorado Springs trip, we had the Business car handled on the rear of a Fort Worth & Denver-Colorado & Southern train which took a short cut to the north. Had we used the Santa Fe, we would have had to make the long detour via Newton, Kansas, or Albuquerque, New Mexico. This was in the days before the I.C.C. put a stop to the free movement of one road's business cars on the rails of another line. Even at Colorado Springs some work still went on. We received mail and telegrams but the work load was light. Colorado Springs was part of the General Manager's territory and before returning to Amarillo we went to Denver for what I guess was business and then returned via Albuquerque. This was something like the President today going to Camp David or his second home. The work always follows."

Cars in the `Sixties

By the 1960s, the widespread use of Business cars was beginning to slow down and eventually disappear, at least in the sense that Ed Mahoney describes above. In the mid-1920s, Santa Fe ordered "shorty" 60 foot steel business cars for use by superintendents. (These are covered in detail in chapter five.) More utilitarian than their predecessors, they were staffed by a cook who doubled as a porter and when traveling the line a secretary (still male) to the superintendent.

Steve Patterson hired out as a stenographer on the Atchison in 1965, and was promoted to secretary to the superintendent at La Junta, Colorado the following year. In an article published in the September/October 1999 issue of *Vintage Rails,* Patterson described in detail his life and adventures aboard a "biz" car. The first car he worked on was number 405 and he remembered it being stocked with the Santa Fe's Bleeding Blue china pattern. His tenure on the car was short, lasting only a few months, before he was reassigned as secretary to general manager of the Eastern Lines, M.L. Olson. Car 36 was assigned to Olson. By the time of Steve's service aboard the 1924-built heavyweight car, it had been modernized with air conditioning, thermopane windows and was painted in the silver shadow-lined scheme. While the car was larger than the 405 by 22 feet, the secretary's room was a compact six by seven feet, smaller than what he previously had, due to the built-in desk. Previously, in the 405, his desk was next to the buffet in the corner of the car's dining room.

Car 36 had both a cook (or chef) and a porter. All, including the secretary, had to work together to make their boss's job easier. Once a trip was announced, Steve's job was to notify the chef and porter so they could order provisions and prepare the car for departure. The trick, of course, was to make sure you had everything you need, since you never could be too sure where the car would wind up.

In each of the cars, there were certain practices and protocols. Dinner was served on fine china, typically California Poppy (and in certain cases Mimbreño) with linen and silver service. Gracious and refined table manners were required. Long-standing practice was that the senior officer/host would sit at the head of the dining table closest to the kitchen (riding backwards), with the second in rank to his right, and so on down the table on both sides. The secretary was seated opposite the boss at the far end of the table, facing forward.

Steve further describes the car's layout that accommodated ten people: "My room was always the farthest from the dining room or the first bedroom after entering the car. Car 36 had three staterooms. Room one, the boss's, was the largest, with a double bed and it was in the center of the car for the best riding qualities. Next was a guest room, which also had a double bed. Between that room and mine was a bathroom with a shower stall made of clear plate glass and framed with stainless steel. After anyone indulged in a shower, Vince (Olson's porter) wiped down and polished the stall to keep it spotless. My berth and the boss's bed were perpendicular to the car and the track." (Compare Patterson's description with that of Mahoney's, above.)

When H.J. "Harry" Briscoe was superintendent of the Eastern Division in the 1960s, he utilized his assigned 400-series car to play an important role in the Santa Fe's safety program. Every quarter, he would take the business car to Chanute, Kansas, from his headquarters in Emporia to assist with the quarterly safety meeting. Being in the field this way underscored management's commitment to safety. Briscoe notes that he averaged about one trip a month on the business car.

As the 'sixties drew to a close, the end was near for most of the Santa Fe's Business cars, victims of the advances in communications. Here, Briscoe relates the end for his car 404, while on a trip with a correspondent who, in turn wrote about it, much to his surprise: "At Emporia, I developed a friendship with Bill White, the son of the famous Kansas editor, William Allen White. Bill was a well-known writer and foreign correspondent in his own right, who had inherited the *Emporia Gazette*. He had become its editor, albeit a non-resident one most of the time as he lived in New York and relied on the managing editor to run the paper under his guidance. During a time when he was in Emporia for a short while, I asked him to be the program at our Rotary Club, which he did and I put up a display in the passenger waiting room of our depot featuring interesting items about William Allen White. Later, I invited him to accompany me on the business car over the Second District through Ottawa, returning over the First District through Lawrence and Topeka. I had the division engineer, Al Ewart, along to give us some professional track information. The train service then was such that we could have lunch leaving Emporia and dinner as we returned through Lawrence and Topeka. It so happened that this was the very last trip for the

The Rose Parade specials consisted of Business cars (the top four lightweights, of course), diners and lounges. In Pasadena, Santa Fe's line crossed Colorado Boulevard, the route of the parade, just north of the depot in downtown. Cars Santa Fe and Topeka were usually spotted on the tracks on opposite sides of the boulevard where the tracks crossed, affording those aboard the best seats in the house to watch the parade. After the parade, the train was spotted at the depot on the house tracks where guests were fed dinner and then bussed to the Rose Bowl for the game. After the game, guests were bussed backed to the waiting train and departure to Los Angeles. The interior of diner 61 Fred Harvey, and view of the special, right, were taken in 1991, while the view upper right was taken at Pasadena New Years Day 1968.
-Upper right Harry D. Peat; others, Mike Martin

While in later years Santa Fe provided hospitality trains, made up from its Business car and Director's car fleet at national conventions for both the Republicans and Democrats, its cars were occasionally leased by aspiring candidates for lesser public office. Here Car 34 is seen at Sweetwater, Texas, in February 1972 while to leased to Ben Barnes for conducting a whistle stop campaign. The 32 year-old Democrat was running for governor of Texas. He lost. *Two photos, Steve Sandifer*

Business car Santa Fe, with Santa Fe President John Reed aboard, along with Car 38 and an unidentified power car, brings up the rear of the *Super C* at Needles, California, on the afternoon of January 15, 1972. While Business cars had been used to entertain shippers for decades, the idea of selling a freight train to potential customers by demonstrating its performance first hand was an innovation which became increasingly common as competition with other forms of transport intensified in the latter part of the twentieth century. *-Phil Kauke*

business car, as we were advised that business cars were being replaces by hy-rail cars (automobiles rigged with steel wheels to could be lowered on the track at street crossing and run on the rails). For five consecutive days, Mr. White wrote a long column about this trip. At the end of the last one, he related, 'But that run was probably the last ever to be made by this private car. For this particular glory has departed, probably forever, from railroading. The Santa Fe has sent an order thundering out of Chicago that, as an economy measure, its division superintendents must give up their private cars. The dance is over. The lights are going out. Poor Harry Briscoe has been whittled down to size and must dead-head around his empire on day coaches like any section hand…And what will happen now that Al Ewart can no longer take his notes? Let them worry about this up in Chicago.' (signed) W.L. White."

Special Trains

The Santa Fe Business car fleet wasn't typically used for special trains, except for director's specials. These were trains where top railroad officials and the board of directors host and entertain top customers to special events, such as the Rose Parade, Rose Bowl (or other game) or quadrennially, the Republican and Democrat Party conventions. Trains of this type became popular in the 1960s and later.

Depending on whether you were working the specials or were a guest aboard, the experience could vary widely. The Rose Parade specials consisted of Business cars (the top four lightweights, of course), diners and lounges. In Pasadena, Santa Fe's line crossed Colorado Boulevard, the route of the parade, just north of the depot in downtown. Cars SANTA FE and TOPEKA were usually spotted on the tracks on opposite sides of the boulevard where the tracks crossed, affording those aboard the best seats in the house to watch the parade. After the parade, the train was spotted at the depot on the house tracks where guests were fed dinner and then bussed to the Rose Bowl for the game. After the game, guests were bussed backed to the waiting train and departure to Los Angeles.

The crew had a somewhat different perspective. Retired Los Angeles Division engineer Don Richardson has written that "the crew was released under the hours of service law as soon as the train was spotted on the house tracks (at the Pasadena depot) after the parade, going on duty whenever the game was over. During that interim, they were fed the same dinner that everyone else had eaten earlier in the diners, then allowed to occupy one of the 'lesser' cars in the executive fleet until going back on duty.

"Actually, it wasn't a very popular job, going on duty in the wee hours after New Year's Eve and spending all of the New Year's Day holiday away from home, hearth and family working for the lowest rate of pay."

Another type of special was run on in November, 1989 on Santa Fe symbol freight *QNYLA* (hot shot intermodal) Chicago to Kansas City with Business cars and then-President Mike Haverty on board with a group of important shippers—JB Hunt. The success of that trip led to a partnership with JB Hunt under the Quantum name and established the Santa Fe as a leader in the intermodal business.

Other Cars, Other Uses

Santa Fe's Business cars were sometimes used for other purposes, even loaned (or "leased") for use by certain important individuals, dignitaries, politicians and even celebrities. One such person, who could be described using all four adjectives, was Dwight David Eisenhower.

The General and Maimie, as they were fondly known, were frequent guests of the Santa Fe in their travels from Gettysburg, Pennsylvania, to either Abilene, Kansas, or Palm Springs, California. Trip planning would start when a memo would go out from the general manager. Contained in the memo would be such information as the daily schedule with departure and arrival times, the equipment to be used (In the 1960s typically one of the named Business cars and a REGAL or PINE series Pullman. The Eisenhowers occupied the Business car and the rest of the party—sometimes 15 to 20 people—the Pullman). If the cars were leaving Harrisburg (the closest station to Gettysburg with rail service) on the Pennsy, the cars were delivered with the observation end of the Business car and the bedroom end of the Pullman pointed east, so they would be headed in the right direction for the trip west. Orders distinctly pointed out that "no publicity should be given this move." The movement on the Santa Fe was to also have a Trainmaster and a Road Foremen on board, riding their respective territories, as well. Additional instructions, such as a clothes rack in the Pullman and a special reading light in the master bedroom of the Business car were included. Also, if the cars laid over at interim points or for a round-trip, then telephones and generators (or a steam generator car) to supply power were to be used. Cars were initially supplied with provisions, but replacements and additions were handled en-route, often by station agents buying the necessary items at a local grocery store, based on wires sent ahead by railroad officials on board. Every step of the party's move was followed by Assistant Vice-Presidents, General Managers, Superintendents, Trainmasters and just about every company official in-between.

With his home town Abilene, Kansas—a Santa Fe town—Ike and Mamie Eisenhower were frequent guests of the Santa Fe. Here they are seen detraining from the car ATCHISON. *-Santa Fe Railway*

When the General and his party would travel to Abilene, local freight trains 73 and 74 would handle the move between Emporia and Abilene. It was left to local officials

Topeka - June 12, 1963

Personal

Messrs. Noe
Lester
Marlo
Danneberg

General Eisenhower and party, occupying a Santa Fe business car and Pullman sleeper, have tentatively planned the following trip:

June 30th: Leave Harrisburg on Pennsylvania Broadway, 9:08 PM.
July 1st: Arrive Chicago 8:00 AM.
Leave on Santa Fe Train 123 11:00 AM.
Arrive Kansas City 8:50 PM.
Leave on Train 1 10:50 PM.
July 2nd: Arrive Emporia 12:30 AM.
Leave on Train 73 2:40 AM.
Arrive Abilene 6:00 AM.
July 3rd: Leave Abilene Train 74 about 3:00 PM.
Arrive Emporia 6:00 PM.
July 4th: Leave Emporia Train 123 12:30 AM.
Arrive La Junta 7:55 AM.
Leave La Junta Trains 201-28 1:05 PM.
Arrive Denver 6:05 PM.

The equipment will probably be released at Denver as the party is expected to return to Harrisburg via another railroad.

It is understood by the party that Trains 73-74 are freight trains handling local work on the Strong City District. It will likewise be satisfactory to handle the party on either the head or rear end, depending on which is more convenient. However, if local work is performed in either direction, it will of course be necessary to handle the party on the rear end behind the caboose. I am hopeful that by a little advance planning we can hold either train down to a maximum of around 15 or 20 cars, as well as local work to a minimum.

- 2 -

Attention is also called to the fact that the party will lay over and occupy the cars from early morning July 2nd to mid-afternoon July 3rd. At least the Pullman car will require steam for air-conditioning, which of course means that we will have to provide a steam generator-equipped diesel unit during the lay over. A Mechanical supervisor should be on the train and at Abilene during the full time the party is on the Eastern Lines. Advance arrangements should also be made for servicing the equipment, including fuel for the diesel unit.

Trainmasters should also ride the train over their respective territories and remain at Abilene during the lay over. Road Foremen of Engines should likewise ride the train over their respective territories.

Mr. Holper will please have a representative of his department on the train and also at Abilene.

All of you should understand these tentative plans should be kept on a confidential basis as no publicity is wanted.

As soon as the trip is definitely firmed up or confirmed, we will receive word about final plans. Meanwhile, you should be laying your plans so that everything will go smoothly. Particular attention should be given the Strong City District between Neva and Abilene.

L. M. Olson

cc: Messrs. Spears
Bill Cox
Holper

An Eisenhower trip on a Santa Fe Business car started out with a letter like this, above, outlining the tentative schedule and details. Note that officials were to be present every step of the way to ensure a smooth trip for the General and Mrs. Eisenhower. The Eisenhower party was aware the trip between Emporia and Abilene was via local freight and that switching might occur enroute. This didn't turn out to be the case as local trainmasters advised crews to do the work at some other time. Above right, in reply to General Manager Olson's letter, J.P. Spears, superintendent of the Middle Division, released a memo detailing the conditions and arrangements made in Abilene, regarding the Eisenhower's trip home to celebrate July 4th in 1963. No detail was overlooked aboard Santa Fe Business cars. Note heater cars, providing steam for the steam ejector air conditioning equipment were used—it can get hot in Kansas in July! *-Keith Jordan collection*

PX

Newton, June 25, 1963

Briscoe - Emporia
Ruegg - Newton
Nail - Newton
Danforth - Newton
Smalley - Newton

Heater car 9005 should be on rear end Train 123 from Chicago on Wednesday, June 26th, to be cut out at Emporia and held at that point to be used on Train No. 73 leaving Emporia Tuesday, July 2nd, with one Pullman and Business Car Santa Fe to supply steam heat for air conditioning. Please arrange to cut it off at Emporia and will advise later how should be lined up on Train 73 leaving Emporia. Briscoe advise arrival. A-76 JT HJB DGR EBN TBD EOS

Spears 1:45 PM

cc: Mr. W. R. Cowdrey - Argentine (Mail)

TRAINGRAM

Topeka, December 9, 1963.

PERSONAL:

Messrs. Elterman
Spears
Bruce
Briscoe
Cenn

General Eisenhower will leave Harrisburg, Pennsylvania on December 15 enroute San Bernardino. Business car Santa Fe and Pullman Standard Pine Leaf will be used and they should be delivered to the Pennsylvania at Chicago before noon December 13, with observation end of business car and bedroom end of Pullman pointing East for deadhead movement to Harrisburg on their No. 50 or No. 54.

These cars will arrive Chicago at 330pm, December 15 on Pennsylvania No. 55 and will deliver to us for movement on Super Chief section train 17, leaving Chicago that evening.

The handling given this party on previous trips has been satisfactory and should be followed closely on this trip. As usual no publicity is to be given this move and any reference to it should indicate the car is in charge of Mr. Stuppi on our line.

Please arrange to have a Trainmaster ride in the rear Pullman with a pakset as on prior trips and have a Road Foreman on the engine throughout the trip.

My information indicates that Mr. Blickle has been requested to see that a clothes rack is provided in the sleeper, as on other trips and special reading light placed in the master bedroom of the business car. Also Mr. Hastings has been instructed to have a representative aboard and I assume Messrs. Blickle and Hastings have handled this matter direct with Messrs. Marlo and Holper.

Please acknowledge by wire, referring to my P-36.

Copy to:
Messrs. Bruce - KCUD
Marlo
Danneberg
Holper

J. E. Lester

Right: The Eisenhowers were no strangers to Santa Fe and rode many trips from Pennsylvania aboard the railroad's business cars. This letter details one such trip (one of at least two in 1963 on the Santa Fe), outlining the requirements and setting in motion a special movement for a special customer. *-Keith Jordan collection*

whether the cars were handled on the head end or the rear end and the train was limited to 15 cars with no regular switching allowed. No detail was too small to generate a wire and a reply, down to the order of the cars on a given track and the speed at which the train should be run, to provide the "smoothest ride."

Sometimes a celebrity would travel on the Santa Fe in their own private car, rather than in a Santa Fe car. In that case, a "Transportation Advice Bulletin" or TAB, would be issued to those involved, with the name, date, equipment, delivery, train position, fare basis, schedule, route, tickets, meals, baggage, and remarks. The Transportation Advice would be wired asking all concerned to "show every courtesy and attention."

ATSF Business car No. 62 brings up the rear of Extra 5034 West as it rolls through Hackberry, Arizona, at speed on the Third Subdivision of the Albuquerque Division on December 28, 1988. *-Mark R. Wayman*

ATSF business cars Santa Fe and Mountainair bring up the rear of the 991 train as it races east at Cameron, California, on April 26, 1986. On the point, three EMD GP60s demonstrator units are providing the power. *-Mark R. Wayman*

Changing Times

Although Mike Martin had photographed a few Business cars on trains, his first opportunity to ride in a business car came in October 1977, one month after being hired by Santa Fe Railway to fill the junior public relations position at the Coast Lines headquarters in Los Angeles. At this time Santa Fe's use of Business cars was in transition.

"In the midst of my final trimester at California State University, Northridge," Mike relates, "I was hired to backfill for Lee Friedersdorf, who was moved to Albuquerque when long-time traveling representative Pat Hill retired from the department.

"One of the tasks on my first Business car trip was to shoot photos and write a story for the monthly employee periodical, *Santa Fe Magazine*. The trip was a shipper special from Hobart Yard in the City of Commerce to the new Barstow classification yard that had opened for business a year earlier. When I arrived at Hobart Yard the train was positioned near the intermodal yard tower, pointed east, with FP45s 5947 and 5945 in freshly washed blue and yellow warbonnet paint. That trip made me aware of the potential for using Business cars to tell the story of the railroad's capabilities with a variety of important audiences, something that would serve the railroad well in future years.

"Santa Fe's Business cars of the late 1970s were something of a link with a past era of passenger travel grandeur. Although some of the furnishings, decor and materials, such as Formica used for counters and desks, and blue and red leather or vinyl chairs, had become badly outdated, the service and the experience aboard Santa Fe Business cars were in the finest tradition of the *Super Chief*. Meals were served on linen table cloths, using authentic Mimbreño pattern dining car china, Albany pattern silverware and glassware etched with the familiar script 'Santa Fe.' Floor ashtrays that doubled as drink glass holders were still in abundance in the lounge and dome-lounge cars, still adorned with their blue and white enamel Santa Fe circle and cross logos at the base. Plastic swizzle sticks proclaimed 'Santa Fe, The Chief Way' and bore a molded Indian chief head with full warbonnet headdress.

"The majority of meals were still heavy fare, with dinners more often than not consisting of steak and potatoes, lunch often the same and breakfasts included Santa Fe French

Toast, eggs, bacon, sausage, biscuits, fruit, all in quantities appropriate for a professional football team. More than once I was told by railroad executives that the greatest danger traveling by Business car was death from overeating! Many of the car attendants and cooks had worked on Santa Fe's passenger trains prior to the 1971 takeover by Amtrak, and they knew how to meet or exceed the expectations of guests and high standards of service for all of those who rode in the company's private fleet.

"During my 20 years with Santa Fe and BNSF, the Business car fleet underwent a transformation beginning in the late 1980s. While management understood the value of Business cars, the fleet was getting old and the cost to operate and maintain it was not insignificant. A complete, business-oriented analysis was undertaken, with all options considered, ranging from retiring the fleet to completely overhauling the cars. Mike Haverty, vice president of operations led the effort, which included Tom Fitzgerald, who was in charge of the freight traffic department. Both understood the costs and value of the Business cars.

"Ultimately, it was determined the fleet, made up of Regal series sleepers, a dome-lounge, low-level lounge, diner, track inspection car, several Business cars and a couple of baggage cars, would be overhauled and the interiors remodeled. The first car to undergo the transformation was the business car Atchison (No. 52). In June 1990, not long after the car was released back into service from it's overhaul and remodeling, I had an opportunity to host Actor Michael Gross and his family on the Atchison from Los Angeles to Chicago coupled to the rear of Amtrak train No. 4, the *Southwest Chief*. The trip was payment-in-kind for Michael appearing in company training videos.

"The late Homer C. Henry, System Road Foreman of Engines, and a good friend to both Michael Gross and I, also accompanied us on the trip, which would feature a return on the rear of the 198 intermodal freight train from Chicago to Los Angeles. On the return trip, we traversed the southern freight route through Amarillo, Texas, and Belen, New Mexico.

"This trip was like many described to me by retired Santa Fe Vice President of Public Relations Bill Burk, whereby Santa Fe invited movie stars, newspaper editors and publishers, titans of business, captains of industry and politicians to ride for a variety of reasons, almost all of whom ultimately formed a favorable perception of the railroad and its management.

"The 1990 trip was nothing short of spectacular, with excellent weather, fine food, good companionship and an opportunity to see the entire northern and southern portions of the lines Santa Fe considered its 'transcontinental' routes. On the eastbound leg of the trip we picked up longtime Santa Fe Modelers Organization officers and directors Chuck Hitchcock and Keith Jordan for the ride to Marceline, Missouri. There was Santa Fe French Toast involved, as well as some pretty healthy railroad-oriented conversation amongst Homer, Michael, Keith, Chuck and I, while we sat in the Business office at the rear of the car and observed the countryside pass by at a speed of 90 miles-per-hour. Perhaps the only downside of this trip was the discharge of disinfected toilet waste from the Amtrak cars ahead of us, a practice that resulted in the Atchison arriving in Chicago coated with a layer of sewage. Thankfully, that practice has since been banned. The Atchison was thoroughly washed in Chicago.

"On the return trip the there was even a birthday party aboard the mile-long intermodal van train after we departed Corwith Yard. It was June 14, and my 35th birthday. Homer had arranged for a birthday cake and the dining area was decorated appropriately. It was the only time I've ever been involved in a Silly String fight aboard a Business car, something revealed in this volume for the first time! As a bonus, we had newly-released GP60M 106 on the point, sporting the updated red and silver warbonnet paint scheme.

"Over time, I was entrusted with more and greater authority by management, especially Coast Lines General Manager Q. W. 'Bing' Torpin, to use Business cars for a variety of community and news media relations activities and image-building occasions. Some included special anniversary trains for which my good friends at Amtrak's west coast public relations office, Arthur Lloyd, then Bruce Heard, helped me facilitate these operations by providing locomotives and former Santa Fe *El Capitan* passenger train high-level cars from the *Sunset Limited*, which did not operate on Saturdays.

"In April 1986 we operated the 'High Desert – Inland Empire Express' to celebrate 100 years of Santa Fe into Southern California and the 10th anniversary of Barstow Yard. The following year I organized the 'Santa Fe Centennial Chief' to celebrate 100 years into Los Angeles on Santa Fe's own rails. That train involved a whistle stop tour from Los Angeles to San Bernardino and return on the railroad's Second District, later known as the Pasadena Subdivision. In addition to a Santa Fe FP45 locomotive leading each trip, and the Amtrak equipment that trailed it, including an Amtrak F40 locomotive supplying head-end power, the rear of each train included Santa Fe lounge car No. 62 and track inspection car No. 89.

"With support from Regional Manager Tom Shalin, in 1991 I was empowered to organize a 'North Pole Express' for children in need of special attention during the holidays. With Tom's authorization to spend some money, Santa Fe employees volunteered to descend upon the local Toys R Us in San Bernardino and buy gifts for the train's special passengers. Manager of Safety Richard Martin capably took on the role of Santa Claus. The first year we visited City of Hope along the Pasadena Subdivision on December 12, taking terminally ill children on a train ride in the Regal Lark (No. 66) and John S. Reed (No. 53), pulled by a pair of blue and yellow SD40-2s (5119 and 5107). We used that same equipment the following day for a trip for Ronald McDonald House children in Orange County, as well as for a group of battered children from the Canyon Acres Home.

"In 1992, our 'Santa Claus Express' was pulled by a more seasonally appropriate red and silver pair of Dash 8-40CW locomotives (846 and 853) when we visited the Valley Division, as well as City of Hope again, using baggage car 77, sleeper Regal Hunt (No. 64), diner Fred Harvey (No. 61), dome-lounge (No. 60) and the William Barstow Strong track inspection car (No. 89). In 1993 the tradition continued with a pair of Dash 8-40BW locomotives (581 and 582) pulling sleeper Regal Crest (No. 63), Fred Harvey diner (No. 61), lounge (No. 62) and the track inspection car (No. 89). To my knowledge, these trips were continued after I left Southern California for the Chicago area headquarters and even after BNSF was formed.

"There were many other occasions when Business cars saw duty, ranging from Operation Lifesaver trains and shipper specials, to Rose Bowl specials, dinners aboard the

Coast Lines business car (No. 58) for winners of a Redlands, California, symphony auction package, Railfairs at Sacramento, inaugural *San Joaquin* Amtrak trains, line sale inspection trips, the 1992 Los Angeles-to-Chicago employee excursion train round trip behind steam locomotive 3751, and others.

"In the late 1980s and early 1990s there were numerous trips from Los Angeles and San Bernardino to Cadiz, California, on the Needles Subdivision, for a proposed joint venture project involving the rail transport and disposal of municipal solid waste from the Los Angeles-Orange County-Inland Empire regions at a landfill in the Mojave Desert. Santa Fe teamed up with Waste Management on this project, dubbed 'Rail-Cycle.' The special trains were operated for news media, government officials, environmentalists and other interested groups, usually sporting a pair of red and silver locomotives and up to six cars from the business car fleet. The wye at Cadiz was used to turn each train for the return trip.

"In general, Business cars were considered by many operating and mechanical people to be a nuisance that diverted resources from other tasks. Certainly, they were disruptive at times, almost always having priority over the functioning of the railroad wherever business cars were being moved or positioned. Yet, there was also an enormous sense of pride felt by employees when they saw those sparkling silver, stainless steel, fluted streamline cars in a string on the rear of a freight train or operated as a special train. Everyone knew they were indeed 'special,' and often carried important people. I was privileged to have been around the cars and ride them throughout the Santa Fe system.

"After the BNSF merger in 1995, my time around the cars was limited to special trains operated for the Galesburg, Illinois, Railroad Days and employee specials that the company inaugurated in June 1997. Jim Schwinkendorf of the operating department worked with me to put together the first BNSF employee appreciation excursion train, a 4,500-mile, month-long trip that kicked off with the dedication of the newly rebuilt Argentine Classification Yard outside of Kansas City. I rode the train from Denver to Laurel, Montana, then to Gillette, Wyoming and eventually to Alliance, Nebraska. The train was comprised of both former Burlington Northern and Santa Fe Business cars, which made up an interesting yet non-uniform looking train.

"The last Business car activities I was involved with was

Mechanical people clean windows on a Director's Special at San Diego, above, in 1991. With decades of superior passenger service as a legacy, Santa Fe's knew how to meet or exceed the expectations of guests aboard its Business car fleet. The addition of Warbonnet-painted Super Fleet power, as with this Rail-Cycle special at Summit, right, reinforced the Santa Fe tradition of high standards service with the *Super Chief* and the *El Capitan* in an intermodal age. -*Two photos, Mike Martin*

A Santa Fe shipper's special moves through the Arizona hill country in April 1990. Even though expensive to maintain and operate, Santa Fe realized that the Business Cars were an excellent public relations tool. They became a primary marketing tool as well, allowing Santa Fe to show off its services in a unique and dynamic way. Customers, government officials, and other VIP's, invited aboard the Business cars considered it very special and never failed to come away from the experience with a favorable impression of the railroad. *-Mike Martin*

some photography of the cars positioned outside the BNSF headquarters building at the company's Fort Worth campus. As I was shooting medium format transparencies on a crystal clear morning, BNSF Chairman Rob Krebs walked by and asked what I thought. 'It is impressive,' I told him, looking at the four Business cars that told any visitor who approached the building that the company had a proud heritage which, at one time, included moving people."

Business Cars as a Symbol

Whether used as a traveling office, hotel, place of entertainment or simply a vehicle to get from one place to another, there is no contemporary to the railroad business car. In his *Vintage Rails* article on life on a Business car, retired Santa Fe employee Steve Patterson summed it up this way, "When an office car came to town, it was the hallmark of power, pedigree and privilege. The car itself spoke *The Company*, and the social implication was exalted status of serenity, aristocracy, and imperviousness, causing the imagination of observers to run wild as to what secrecy occurred on board.

"Those images were scrapped decades ago, when the railroads no longer needed or could afford such gilt-edging for their managers. There is a different business conducted aboard Business cars nowadays. Coupled with an auditorium/ inspection car, there is no better way to show off the railway to shippers, civic leaders, and security analysts. Since railroads don't have static showrooms to display their services, their Business cars become a primary marketing tool to conduct business in a dynamic setting. The cars are excellent for public relations, as clientele are captive to nostalgia, and analysts are able to see the condition of the company, its very flesh and blood, to gain the best understanding how the railroad operates. Customers invited aboard Business cars consider it very special."

Many of the Santa Fe Business cars still serve today, whether as static entertainment/meeting facilities in front of successor BNSF's corporate offices in Texas, or out on the rails on the back of a fast moving freight, with company executives keeping a watchful eye on the railroad.

The opening of the new line to Chicago in 1893 was the occasion for this much-photographed inspection train in Illinois. From the rear, the consist included the six-wheel Business car 215, large-windowed Business car 214, Baggage car 633 and locomotive 748. *-Santa Fe photo, John B. McCall collection*

2

Official Cars Prior to 1903

Most of the early business or official cars on the Santa Fe system were converted from open platform coaches, with a few exceptions, as listed in the accompanying roster. These conversions were crude, as in most cases the rows of seats were simply removed and partitions were erected dividing the rooms from one another. The exterior of the cars remained unchanged for the most part. In some cases a window or two was blanked out, but externally, very little was or needed to be changed.

Available data is insufficient for a comprehensive roster of official cars prior to the year 1885. Data available on these cars between 1885 and 1907 is very vague and much research has been exerted to compile the roster presented for these cars on the following pages.

The first two cars constructed for the Santa Fe to be used strictly in "Official" service came from Pullman in 1893, the Bay State and the Sunflower State. Both of these cars were rebuilt over the years and lasted well into the twentieth century (See chapter 3). With the depression later in 1893 came a slump in business, and no additional official cars were purchased or converted within the next decade.

The official cars of the AT&SF, which remained after the mid-1890s were renumbered in numerical order from AT&SF 212 through 218, as AT&SF 211 was the highest numbered coach at that time. The official cars of the GC&SF retained the numbers listed above until the general renumbering during the early 1900s.

During this period there were many railroads under the control of the Atchison Topeka & Santa Fe by stock ownership, or lease. Many of these properties were eventually merged into the system, but some were sold or traded away and lost all identity with the Santa Fe. A number of private or official cars of these roads eventually came to the Santa Fe and are listed in an accompanying table.

Pay Car ATSF 100 in service about 50 miles west of Dodge City, Kansas, in 1872. Jim Moore, Santa Fe's Paymaster, had this old coach fixed with sleeping quarters and a modest arsenal, loaded aboard with his sacked cash was a half dozen armed guards. The men were paid cash until 1873 when this was decided as to be too dangerous. *-Eagan photo, Denver Public Library, Western History Department*

For example, Business 100 was one of these cars. Constructed by the Atlantic & Pacific Railroad in 1883, it became Santa Fe Pacific 100 in 1897, it was reequipped with six-wheel trucks, and became AT&SF Business 10 in 1902, where it was assigned to the Superintendent of Transportation. Later it was reassigned to the General Manager, SFP&P Lines, and during the 1920s it was assigned to the Division Superintendent of the Arizona Division where it was renumbered AT&SF 2nd 413.

For those cars which survived into the twentieth century, the reader is directed to Chapter Three, where detailed histories, and illustrations of these cars, if available, are presented.

SANTA FE OFFICIAL CARS PRIOR TO 1903

Car No.	*Type*	*Builder & Date*	*Length Trucks*	*Notes*
AT&SF "A"	Official car			Unknown origin prior to 1890; out in 1891.
GC&SF 9	Officer's Inspection car			From GC&SF 9 Mail & Express after 1890; out prior to 1902.
AT&SF 20	Pay car			From Coach AT&SF 20 after 1890; out in 1892.
AT&SF 26	Bagg. & Sleeper (Official)			From Baggage-Mail & Express AT&SF 26 prior to 6/1885; out between 1885-90.
AT&SF 49	Pay car			From Coach AT&SF 49 prior 1890; out in 1891.
AT&SF 63	Official car		4wh	From Coach AT&SF 63 prior to 6/1885; renumbered AT&SF 98 Business, 1888-90; renumbered 214, 10/1893.
AT&SF 97	Official car	AT&SF 1880	41' 4wh	From Coach AT&SF 97 after 1890[1]; renumbered AT&SF 212 Business, 12/1893.
AT&SF 97	Official car	AT&SF 1887	46' 4wh	From Coach AT&SF 97 after 1890[1]; renumbered AT&SF 213 Business, 1/1896.
AT&SF 99	Official car	AT&SF 1887	60' 6wh	From Coach AT&SF 99 prior to 6/1885; renumbered AT&SF 216 Business, June 1894.
AT&SF 100	Pay car	CB&Q 1874	45'	From Coach AT&SF 100 prior to 6/1885; out in 1886.
AT&SF 2nd 100	Official car	B&S 1886	60' 6wh	Renumbered AT&SF 217 Business, January, 1895.
GC&SF 100	Pay car			Unknown origin prior to 1890; out in 1897.
AT&SF 101	Official car	AT&SF 1887	60' 6wh	From Coach AT&SF 101 prior to 6/1885; renumbered AT&SF 218 Business, June 1894.
GC&SF 102	General Manager's car	B&S 1883	52' 6wh	Unknown origin prior to 1890; renumbered GC&SF 13 Business, 1904, renumbered GC&SF 2nd 11, 1905.
GC&SF 103	Official Inspection car		45' 4wh	From GC&SF 5 Baggage-Mail & Express after 1890; renumbered GC&SF 418 Superintendent's Business, 1905.
GC&SF 104	Supt's Business car		40' 4wh	From Coach GC&SF 6 Coach, 1901; renumbered GC&SF 416 Superintendent's Business, 1900.
GC&SF 105	Supt's Business car	1886	45' 4wh	From Coach GC&SF 15 Coach, 1902; renumbered GC&SF 415 Superintendent's Business car, 1904.
AT&SF 137	Pay car			From Coach AT&SF 137 prior to 6/1885; out in 1891.
AT&SF 199	Official car	AT&SF 1887	51' 6wh	From Coach AT&SF 199 after 1890; renumbered AT&SF 215 Business, 10/1893.
AT&SF 201	Official car			From Coach AT&SF 201 prior to 1890; out in 1892.
AT&SF 200	Pay car			From Coach AT&SF 200 after 1890; out in 1892.
AT&SF 300	Official car			From Coach AT&SF 306 after 1890; out in 1896.
AT&SF 500	Coach & Bagg (Official)			From AT&SF 500 Commissary after 1890; renumbered AT&SF 599 Coach & Bagg (Official); out in 1893.
AT&SF 209	Business	CB&Q 1870	44' 4wh	From unknown, 1901[2]; renumbered Superintendents Business AT&SF 402, 1903.
AT&SF 210	Business	CB&Q 1870	44' 4wh	From unknown, 1901[2]; renumbered Superintendents Business AT&SF 407, 1905.
AT&SF 211	Business	CB&Q 1871	45' 4wh	Renumbered from AT&SF 300, 1/00; renumbered Superintendents Business AT&SF 403, 1903.
AT&SF 212	Business	AT&SF 1880	41' 4wh	Renumbered from AT&SF 97, 12/1893; renumbered Superintendents Business AT&SF 410, 1903.
AT&SF 213	Business	AT&SF 1887	46' 4wh	Renumbered from AT&SF 97, 1/1896; renumbered Superintendents Business AT&SF 400, 1903.
AT&SF 214	Business		4wh	Renumbered from AT&SF 98, 10/1893; renumbered Superintendents Business AT&SF 408, 1903.
AT&SF 215	Business	AT&SF 1887	51' 6wh	Renumbered from AT&SF 199, 10/1893; renumbered Business car AT&SF 8, 1904.
AT&SF 216	Business	AT&SF 1887	60' 6wh	Renumbered from AT&SF 99, 6/1894; renumbered Business car AT&SF 7, 1905.
AT&SF 217	Business	B&S 1886	60' 6wh	Renumbered from AT&SF 100, 1/1895; renumbered Business car GC&SF 12, 1903.
AT&SF 218	Business	AT&SF 1887	60' 6wh	Renumbered from AT&SF 101, 6/1894; renumbered Business car AT&SF 6, 1903.
AT&SF 219	Business	Pullman 1893	69' 6wh	New, also named SUNFLOWER STATE; renumbered Business car AT&SF 2, 1904; reno. AT&SF 2nd 4, 1905.
AT&SF 220	Business	Pullman 1893	69' 6wh	New, also named BAY STATE; renumbered Business car AT&SF 1, 1903.
AT&SF 221	Business	B&S 1890	65' 6wh	Rebuilt from DinerAT&SF 1402, 1903; renumbered Business car AT&SF 4, 1903; reno. AT&SF 2nd 2, 1905.
AT&SF 222	Business	B&S 1890	65' 6wh	Rebuilt from Diner AT&SF 1403, 1903; renumbered Business car AT&SF 3, 1905.
AT&SF 223	Business	B&S 1891	65' 6wh	Rebuilt from Diner AT&SF 1408, 1903; renumbered Business car AT&SF 5, 1903.

1. Official records listed two Coaches numbered AT&SF 97 during the early 1890s, both of which were converted to official cars carrying the number 97.
2. Due to the short length of these two cars they are believed to have been converted from old coaches.

Views of early Santa Fe Official cars are exceedingly rare. This car, left, is believed to be the former ATSF Official car 300, built from AT&SF Coach 300 some time after 1890 and off the roster by 1896. Sold to parties unknown, it still appeared to be in service at a rustic location at a later date. *-Frank Ellington collection*

The view below is Santa Fe Official cars 214 and 215, in use on an Illinois inspection trip when brand new. The rear car, AT&SF 215, was converted from Coach 199 to Business 199 in 1891, and renumbered Business 215 in October 1893. In 1904, it was again renumbered, this time to Business number 8, and assigned to the General Superintendent-Western Lines. *-Santa Fe photo, John B. McCall collection*

At left is the opposite side of Official car 214 on the same inspection trip. Originally numbered Coach 63, this car was converted to Business car 63 prior to 1885. It was renumbered to Business 98 about 1889, again renumbered to 214 in October 1893, and finally to Division Superintendent's car 408 in 1903, and assigned to the superintendent of the San Francisco and San Joaquin Valley Railway. As pictured here, the body was deep maroon (high gloss), the lettering and the pin striping was gold leaf with bright red shadow box effects. The roof was black, the boxes and underbody were black and the trucks were a lighter shade of Pullman green. The car had patented paper center wheels on the four wheel trucks. So far as is known, of the first generation Division Superintendent's Business cars, this was the only one with the large picture windows throughout. The only difference from one side to the other was the smaller window on this side, where there is another standard width window on the other side. The car had an observation room on the near end, a couple of staterooms in the middle, a galley (by the narrow window) and a dining room on the other end. Both ends of the car had those deep observation windows onto the platform. *-Santa Fe photo, John B. McCall collection*

James Hagerman, promoter, and Charles Eddy, a Pecos rancher, formed the Pecos Valley Railway in 1890, spent millions of dollars developing the area and railroad, and purchased this private car from the Pullman Company in 1892. The area and railroad were devastated by the depression of 1893 and floods in the valley. Hagerman, aided by the Santa Fe, in 1895 rebuilt the original portion of the line and built north to meet the Santa Fe at Texico. History does not record the car Hesperia passing to the Santa Fe, leading to the possible conclusion that Eddy took ownership and used it as his private car, although there is considerable speculation that it may have become AT&SF Business car 15 (see page 52). *-Pullman photos, courtesy Smithsonian Institution*

OFFICIAL CARS, OTHERWISE UNACCOUNTED FOR

Colorado Midland	99	Officers car	AT&SF control from 11/1890, sold 1896.
Colorado Midland	101	Officers car	AT&SF control from 11/1890, sold 1896.
Sonora Ry.	104	Pay car	Traded to Southern Pacific, 1897.
Sonora Ry.	252	Officers car	Traded to Southern Pacific, 1897.
Pecos Valley RR.	77	Officers car	Named Hesperia; disposition unknown (see page 52).
S.F. Prescott & Phoenix	98	Officers car	Converted to AT&SF 199085 Bunk car, 1924.
Oklahoma Central	400	Officers car	Dismantled 1917 by the AT&SF.
S.F. Rocky Mountain & Pacific	200	Officers car	Converted to AT&SF 199327 Bunk car, 1919.
Northwestern Pacific	04	Officers car	50% SP ownership; converted to NWPMW 340, Bunk car, 12/26/46.
Kansas City Mexico & Orient	50	Officers car	Converted to AT&SF 199178 Bunk car, October 1930.
Kansas City Mexico & Orient	100	Officers car	Purchased with the KCM&O In 1929, disposition unknown.
Kansas City Mexico & Orient	200	Officers car	Purchased with the KCM&O in 1929, disposition unknown.
Kansas City Mexico & Orient	300	Officers car	Purchased with the KCM&O in 1929, disposition unknown.

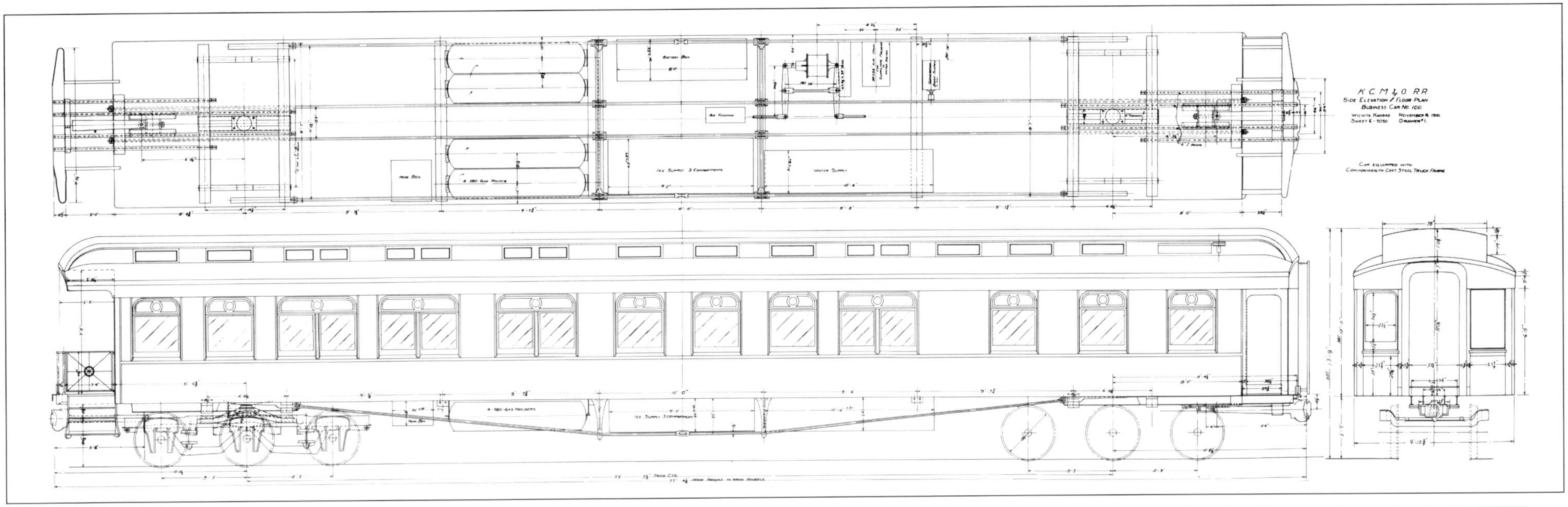

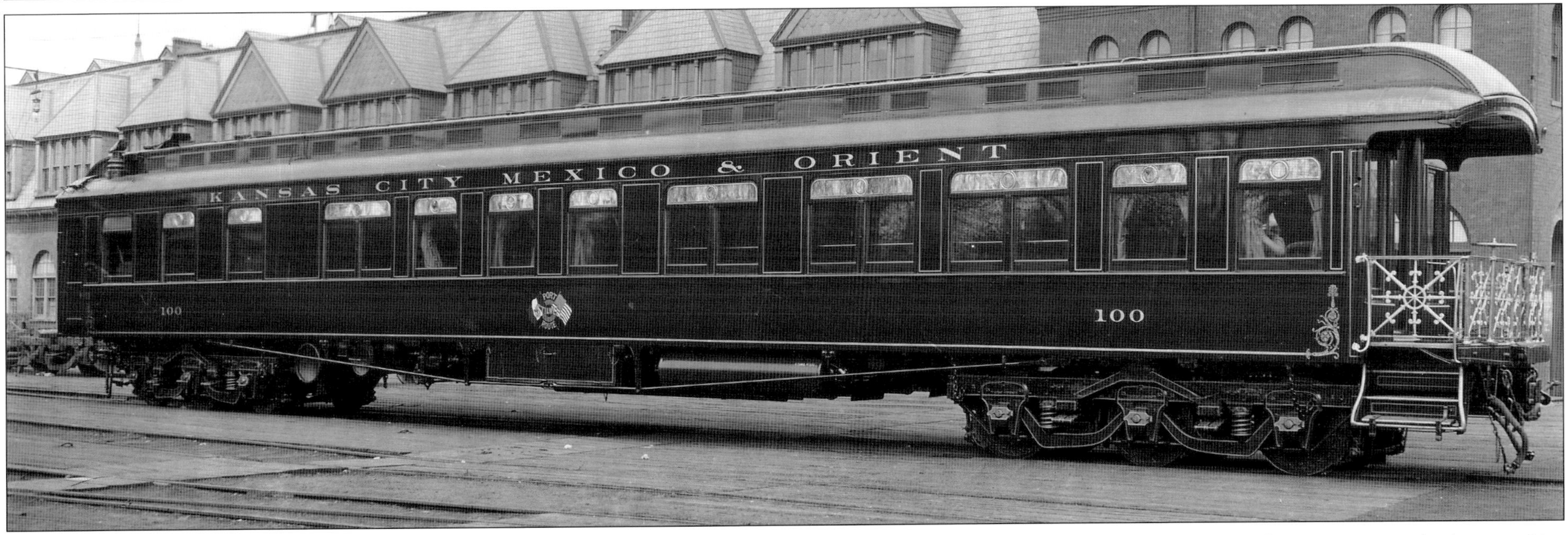

Pullman constructed this car in 1897 for promoter Arthur Stilwell, then in command of the Kansas City, Pittsburg & Gulf. Stilwell lost control of the KCP&G in 1899, but car 100 remained in his ownership. His new venture was to build a shortcut road from Kansas City to the Pacific through Mexico to the port of Topolobampo. Here, in 1903, his car 100 has been relettered for this new venture. "Port Stilwell Route" was embossed over the twin flags of the two countries it planned to serve. A religious man, Stilwell's private car was equipped complete with a parlor organ and pulpit. This car presumably became Santa Fe property with the purchase of the KCM&O in 1929, but its disposition is unknown. *-Pullman photo courtesy Smithsonian Institution*

The Topeka shops converted former Barney & Smith 1890 Diner 1408 into Business 223 in 1903. Later that same year the 223 was renumbered Business car No. 5. Here we find the General Manager, Coast Lines, with his Car 5 "spotted" at some lonely desert location. By 1928, Car No. 5 was assigned to the Assistant General Manager Western Lines (Northern District). Notice the believed to be one-of-a-kind observation platform railing. All known data seems to indicate that railing was applied at the time of the reconstruction into Business car configuration. *-Santa Fe photo, John B. McCall collection*

3

New & Rebuilt Wood Business Cars

During 1903, and continuing through 1914, Santa Fe gathered together official cars from around the system, and from various subsidiaries, and began to develop its first listing of wooden Business cars in the series AT&SF Nos. 1 through 25. They did not all arrive in sequence, however, as Nos. 1 through 10 and No. 16 were renumbered or converted between 1903 and 1906, Nos. 11 through 15, 17 through 19, and 25 came in 1907, Nos. 20 through 22 were added in 1910, No. 23 came in 1913, and No. 24 was converted in 1914.

The first two cars constructed for the Santa Fe to be used strictly in "Official" service came from Pullman in 1893. These were ATSF 220, also named Bay State, and AT&SF 219, also named Sunflower State. In 1903, these became AT&SF 1 and 2, respectively.

With the depression later in 1893 came a slump in business, and no additional official cars were purchased or converted within the next decade. The following decade, 1903-1913, saw the early Dining cars, mostly built by Barney & Smith in 1890, being replaced with new equipment. These displaced Dining cars were naturals for conversion to Business cars as no expensive kitchen section needed to be built into them. The vestibuled end of these old Diners was simply rebuilt into an open platform. During this same period there were also a few Coaches and a Chair car converted for official use. Car Nos. 3, 4, 5, 6, 7, 8, 2nd 12, 16, 18, 21, 22, 23, 24 and 2nd 24 all had origins as Diners or other types of cars.

Two additional elegant private officers' cars were delivered by Pullman in 1907, one for Santa Fe president E.P. Ripley, AT&SF No. 17, and one for a member of the board of directors, AT&SF No. 25.

These wooden cars served the officials of the Santa Fe exclusively through the late 1920s. The cars remaining in service at that time were given a new lease on life when a program was instituted to refurbish and upgrade them to match the new heavyweight all-steel cars being purchased. This program included the installation of a steel underframe and bulkheads and newer trucks. Their wooden exteriors were sheathed with steel plates at this time, and all but the 15 had any gusset windows exposed after rebuilding.

For some reason, there was considerable shifting of numbers within the series prior to 1910: car No. 4 became 2nd No. 2, No. 4 became 2nd No. 2, car No. 13 became 2nd No. 11, car No. 12 became No. 20 and car 1st No. 17 became 2nd No. 16. Then, toward the end of the wood car era, Topeka Shops renumbered AT&SF 2nd No. 4 to 199020 in May 1952 to make room for the heavyweight Business car of the same number, which had just been converted from Coach 3346. Similarly, car No. 22 was renumbered to 199075 in July 1952 to make room for the heavyweight Business car of the same number then being converted from a Coach.

Ancient car No. 20 was likely the first "big" business car Santa Fe ever acquired, a 60-foot car having been built by Barney & Smith as Gulf, Colorado & Santa Fe Business car No. 12 in 1886. Until the heavyweight Business cars began arriving in 1918, the biggest Business cars were the two 69-foot cars delivered in the 1890s, Nos. 219 and 220, and 72-foot cars Nos. 17 and 25, built in 1907.

At least 13 of these steel-sheathed cars were converted for Company service during the 1950s and 1960s, with most of the conversions taking place at Cleburne, Texas.

One car, AT&SF No. 5, originally Barney & Smith 1890-built Diner 1408, wound up its days as Bunk car 199191, but was sold to the Smoky Hill Railway Historical Society in November 1971 and is preserved.

Division Superintendent C. T. McLellan (standing), and party ready for an inspection of the Eastern Division about 1910. They are sitting on an inspection platform similar to the one used in the photo of the 401. *-Santa Fe Collection, Kansas State Historical Society*

SANTA FE NEW & REBUILT WOOD BUSINESS CARS

Business Car Nos.	*Builder Date Built*	*Length & Trucks*	*Notes & Disposition*
AT&SF 1	Pullman 1893	69' 6 wh.	Originally AT&SF 220, Bay State; reno. AT&SF 1, 1903; dismantled 10/28/22.
AT&SF 2	Pullman 1893	69' 6 wh.	Originally AT&SF 219, Sunflower State; reno. AT&SF 2, 1904; reno. AT&SF (2nd) 4, 1905.
AT&SF 2nd 2	B&S 1890	65' 6 wh.	Dismantled 12/28.
AT&SF 3	B&S 1890	65' 6 wh.	Converted to AT&SF 222, 1903, from AT&SF 1403 Diner; reno. AT&SF 3, 1905; sold Amusement Corp. of America, Ft. Worth 11/41.
AT&SF 4	B&S 1890	65' 6 wh.	Converted to AT&SF 221, 1903, from AT&SF 1402 Diner; reno. AT&SF 4, 1903; reno. AT&SF (2nd) 2, 1905.
AT&SF 2nd 4	Pullman 1893	69' 6 wh.	Renumbered from AT&SF 2, 1905; converted to 199020 Bunk car (5/52), (see note A).
AT&SF 5	B&S 1890	65' 6 wh.	Converted to AT&SF 223, 1903, from AT&SF 1408 Diner; reno. AT&SF 5, 1903; converted to 199191 Bunk car 2/12/51; to Smokey Hill Ry. Hist. Soc. 11/71.
AT&SF 6	AT&SF 1887	60' 6 wh.	Converted to AT&SF 101 from AT&SF 101 Coach; reno. AT&SF 218, 6/1894; reno. AT&SF 6, 1903; converted to 199180 Bunk car 1/36.
AT&SF 7	AT&SF 1882	57' 6 wh.	Converted to AT&SF 99 from AT&SF 99 Coach; reno. AT&SF 216, 6/1894; reno. AT&SF 7, 1905; sold GC&SF 3/30/31, dismantled 4/11/32.
AT&SF 8	AT&SF 1887	51' 6 wh.	Converted to AT&SF 199 from AT&SF 199 Coach; reno. AT&SF 215, 10/1893; reno. AT&SF 8, 1904; dismantled 3/11/18.
AT&SF 2nd 8	AT&SF 1899	60' 6 wh.	Converted to AT&SF 2nd 8, 1918, from AT&SF 1028 Chair car; converted to 199146 Bunk car 9/30/53, sold for scrap 5/71.
AT&SF 9	H&H 1882	51' 6 wh.	Reno. SFP 99, 6/1897, from Atlantic & Pacific 99; reno. AT&SF 9, July 1902; reno. AT&SF 2nd 412, Superintendent's car 4/27.
AT&SF 10	A&PRR 1883	56' 6 wh	Reno. SFP 100, 6/1897, from Atlantic & Pacific 100; reno. AT&SF 10, July 1902; reno. AT&SF 2nd 413, Superintendent's Car 9/27.
GC&SF 11	B&S 1890	50' 6 wh.	Disposition unknown, dropped from roster in 1905.
GC&SF 2nd 11	B&S 1883	52' 6 wh.	Originally GC&SF102 Gen. Mgr's car; reno. GC&SF 13, 1904; reno. GC&SF 2nd 11, 1905; reno. AT&SF 11 5/27/31; dism. 12/23/31.
GC&SF 12	B&S 1886	60' 6 wh.	Originally AT&SF 100 Business; reno. AT&SF 217, 1/1895; reno. GC&SF 12, 1903; reno. AT&SF 20, 1910.
GC&SF 2nd 12	B&S 1890	65' 6 wh.	Converted from AT&SF Diner 1408, 1910; Reno. AT&SF 12 5/27/31; conv. to 199199 Shovel Operators car 2/53; sold for scrap 4/72.
GC&SF 13	B&S 1883	52' 6 wh.	Originally GC&SF 102 General Manager's car; reno. GC&SF 13, 1904; reno. GC&SF 2nd 11, 1905.
AT&SF 14	Pullman 1881	60' 6 wh.	Originally Southern California Ry. Officers' car 98; reno. SFP&P 14, 1904; reno. AT&SF 14, 1912; dismantled 3/7/19.
AT&SF 15	1907	62' 6 wh	Originally E Ry. of NM Officer's car 15; reno. AT&SF 15, 1/12; converted to 199148 Bunk car, 10/53; retired and sold for scrap, 2/72.
AT&SF 16	St.LCCo 1886	56' 6 wh	Converted to AT&SF 16 Business from ATSF 2697 Coach 1907; dismantled 8/28/37.
AT&SF 2nd 16	Pullman 1907	72' 6 wh	Reno. from AT&SF 17, 11/52; converted to 194020 Bunk car 6/60; destroyed by fire 6/64.
AT&SF 17	Pullman 1907	72' 6 wh	Purchased new, reno. 2nd AT&SF 16, 11/52.
AT&SF 18	B&S 1890	65' 6 wh	Converted to AT&SF 18 Business, 1907, from AT&SF 1404 Diner; converted to 194021 Bunk car 12/61; sold for scrap 6/71.
AT&SF 19			Originally PV&NE 14 Business; reno. PV&NE 1400, 7/1899; reno. PV&NE 15, 7/04; reno. AT&SF 19, 4/08; dismantled 12/30/35.
AT&SF 20	B&S 1886	60' 6 wh	Originally GC&SF 12; reno. AT&SF 20, 1910; converted to AT&SF 20 Track Recording Car, 7/27.
AT&SF 21	B&S 1890	65' 6 wh	Converted to AT&SF 21 Business, from AT&SF 1407 Diner, 1910 ; converted to AT&SF 21 Radio Test Car 1940s est.
AT&SF 22	B&S 1890	65' 6 wh	Converted to AT&SF 22 Business, from AT&SF 1405 Diner, 1910; converted to 199075 Bunk car 7/52; sold for scrap 12/72 (see note B).
AT&SF 23	Pullman 1888	65' 6 wh	Converted to AT&SF 23 Business, from AT&SF 1400 Diner, 1913; assigned as a "Demonstration Car" circa 1934.
AT&SF 24	B&S 1885	45' 4 wh	Converted to AT&SF 24 Business, from GC&SF 2810 Coach, 1914; renumbered ATSF 199211, Corwith Pay Car 5/15.
AT&SF 2nd 24	B&S 1900	70' 6 wh	Steam Heat Instruction Car from AT&SF 1412 Diner, 10/14; converted ATSF 789, Coach Smoker 1/26 (see Special Purpose cars, Chapter 8.)
AT&SF 25	Pullman 1907	72' 6 wh	Purchased new; sold to Texas Delta Farms Co. 1/17/20.

Key: AT&SF- Atchison Topeka & Santa Fe; GC&SF - Gulf Colorado & Santa Fe; E Ry. of NM - Eastern Railway of New Mexico; PV&NE - Pecos Valley & Northeastern; SFP - Santa Fe Pacific (ex-Atlantic & Pacific)

Note A: *Topeka Shops renumbered 2nd AT&SF 4 to 199020 in May 1952 to make room for the heavyweight Business car which had just been converted from Coach 3346. The Business car numbered 199020 was then converted to a Foreman car 199020 at Cleburne, Texas, in September 1953.*

Note B: *Topeka Shops renumbered Business 22 to 199075 in July 1952 to make room for the heavyweight Business car which had just been converted from Coach 3360. The Business car numbered 199075 was then converted to a Foreman car 199075 at Cleburne, Texas, in May 1953.*

WOOD BUSINESS CAR ASSIGNMENTS

Business Car No.	*Assignment 1907*	*Assignment 1910*	*Assignment 1919*	*Assignment 1928*
1	Directors	Directors	Unassigned	-
2	Unassigned	-	-	-
2nd 2	-	General Manager Western Lines	Unassigned	Unassigned
3	Vice President Traffic	Vice President Traffic	Vice President Traffic	Unassigned
4	Vice President Operations	-	-	-
2nd 4	-	Vice President Operations	General Manager Western Lines	Asst. to Vice President Mechanical Dept.
5	General Manager Coast Lines	General Manager Coast Lines	General Manager Coast Lines	Asst. General Manager Western Lines (Northern Dist.)
6	Unassigned	General Superintendent Coast Lines	Assistant General Manager Coast Lines	Superintendent Plains Division
7	General Supt. Eastern Lines (Eastern Dist.)	General Supt. Eastern Lines (Eastern Dist.)	Asst. General Manager Eastern Lines (Eastern Dist.)	Asst. General Manager Eastern Lines (Western Dist.)
8	General Superintendent Western Lines	General Superintendent Western Lines	-	-
2nd 8	-	-	General Manager Eastern Lines (Western Dist.)	Assistant General Manager Eastern Lines
9	General Purchasing Agnt	Unassigned	General Storekeeper	-
2nd 9	-	-	-	Unassigned
10	General Superintendent Coast Lines	Superintendent of Transportation	General Manager SFP&P	-
2nd 10	-	-	-	Unassigned
2nd 11	General Superintendent Gulf Lines	General Superintendent Gulf Lines	General Superintendent Gulf Lines	Superintendent Southern Division
12	General Manager Gulf Lines	-	-	-
2nd 12	-	Vice President & General Manager Gulf Lines	Vice President & General Manager Gulf Lines	Assistant General Manager Gulf Lines
13	-	-	-	-
14	President & General Manager SFP&P	President SFP&P	-	-
15	President & General Manager Pecos Lines	Superintendent ERyNM	General Supt. Western Lines (Southern Dist.)	Asst. General Manager Western Lines (Southern Dist.)
16	Superintendent Motive Power	Superintendent Motive Power	Asst. to Vice President, Chief Mechanical	Superintendent Middle Division
17	President E.P. Ripley	President E.P. Ripley	Vice President Operations	Unassigned
18	General Manager Eastern Lines	General Manager Eastern Lines	General Manager Eastern Lines	Chief Engineer System
19	General Supt. Eastern Lines (Western Dist.)	General Supt. Eastern Lines (Western Dist.)	Superintendent New Mexico Division	Superintendent Slaton Division
20	-	Chief Engineer	Chief Engineer System	Track Recording car
21	-	Vice President Purchasing Dept.	Vice President	Purchasing Agent
22	-	Vice President Engineer Dept.	General Supt. Western Lines (Northern Dist.)	General Manager Western Lines
23	-	-	Unassigned	Unassigned
24	-	-	Steam Heat Instruction	-
25	B.P. Cheney (Director)	B.P. Cheney (Director)	-	-

1
A.T.&S.F.

Santa Fe Business car No. 1, built by Pullman in 1893, began life as AT&SF 220, Bay State, and was renumbered AT&SF No. 1 in 1903. This car, and sister 219, Sunflower State, were unquestionably the most unique Business cars on the system, inasmuch as they were constructed without the traditional open platform. Bay State and Sunflower State were built during an era when car building was considered an art, and they were masterpieces with their glass paneled vestibule doors, large end windows and intricate pin striping, especially on the truck sideframes. Car No. 1 had six single, and two double beds and a sleeping capacity of ten. The car as built had six wheel trucks, a light weight of 139,860 lbs. and a 6-inch "I" beam underframe with steel platforms. It was dismantled October 28, 1922. *-Pullman photo courtesy Smithsonian Institution*

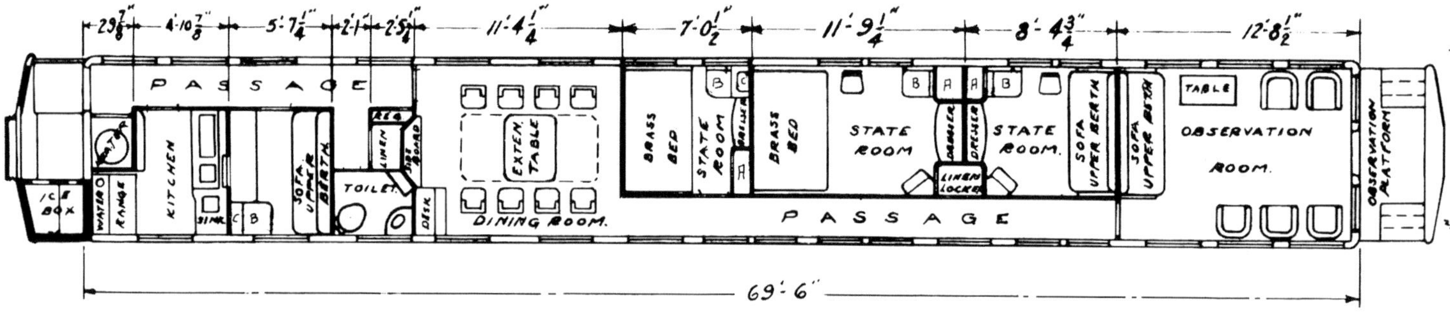

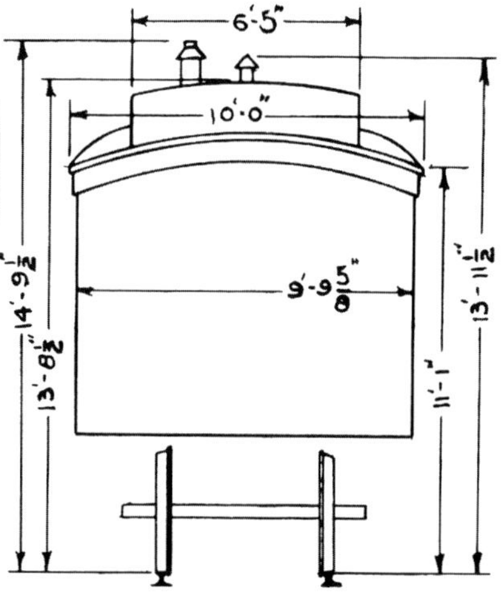

Business Car No. 1
(former No. 220, Bay State)
Pullman 1893

Many of the early presidents and directors of the Santa Fe maintained their executive headquarters in Boston, hence the name Bay State for this Director's car.

Rare in service view of Business car 220, the Bay State, "out west" somewhere in the 1890s, prior to being renumbered AT&SF No. 1 in 1903. *-John B. McCall collection*

2 A.T.&S.F.

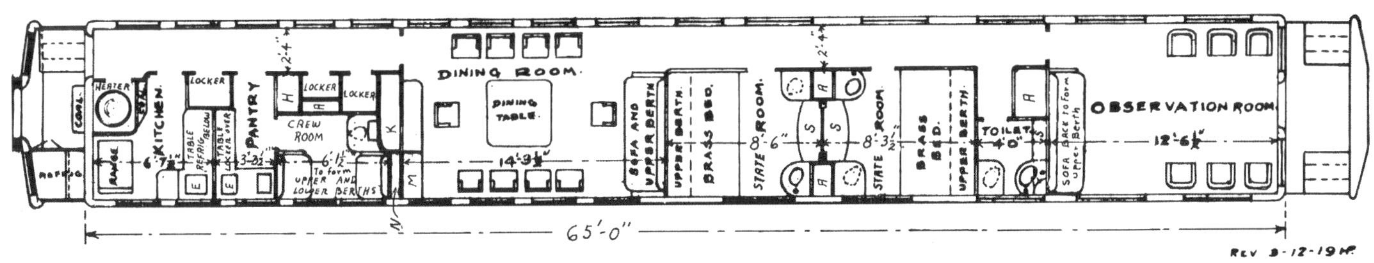

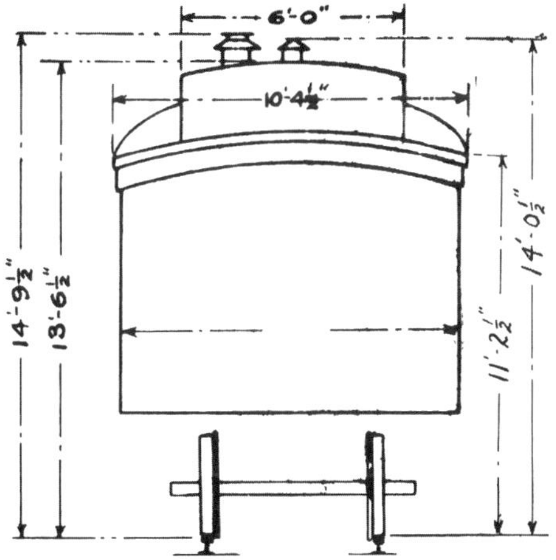

Business Car 2nd No. 2
(Rebuilt from Diner 1402, former 221, former 4)
Barney & Smith 1890

AT&SF Business car 2nd No. 2, started life as a 65-foot Diner built by Barney & Smith for the Santa Fe in 1890 and numbered 5 (renumbered 1402 by 1900). The six-wheel car was converted to Business car 221 and renumbered AT&SF 4 in 1903. Two years later it was renumbered 2nd 2, the number it carried until being dismantled in December 1928. The car had six single and two double beds with a sleeping capacity of 12. It was built with a 6-inch "I" beam underframe and steel platforms and had a light weight of 131,500 lbs. To date, no photos of 2nd No. 2 have surfaced.

AT&SF Business car No. 3, started life as a 65-foot Diner built by Barney & Smith for the Santa Fe in 1890 and numbered 6 (renumbered 1403 by 1900). The six-wheel car was converted to Business car 222 and renumbered in 1903, and renumbered AT&SF No. 3 two years later. The car had six single and three double beds with a sleeping capacity of 12. It was built with a 6-inch "I" beam underframe and steel platforms and had a light weight of 137,500 lbs. Steel underframe and steel plating were added to the car during the mid-1920s, just prior to this photo being taken at Topeka, Kansas, July 23, 1928. *-Santa Fe photo, Frank Ellington collection*

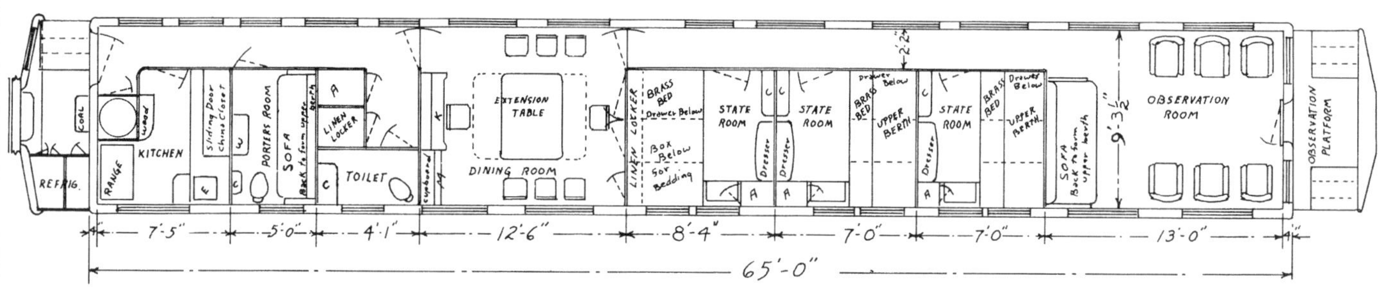

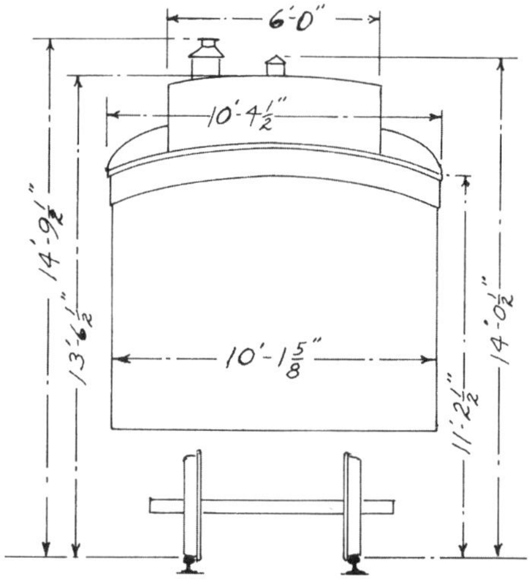

Business Car No. 3
(Rebuilt from Diner 1403, former 222)
Barney & Smith 1890

Opposite side of car No. 3 at Topeka in July 1928. The car could trace its lineage back to Barney & Smith Diner No. 5 of 1890. The early Diners were often chosen for conversion to Business cars because of the galley facilities. *-Frank Ellington collection*

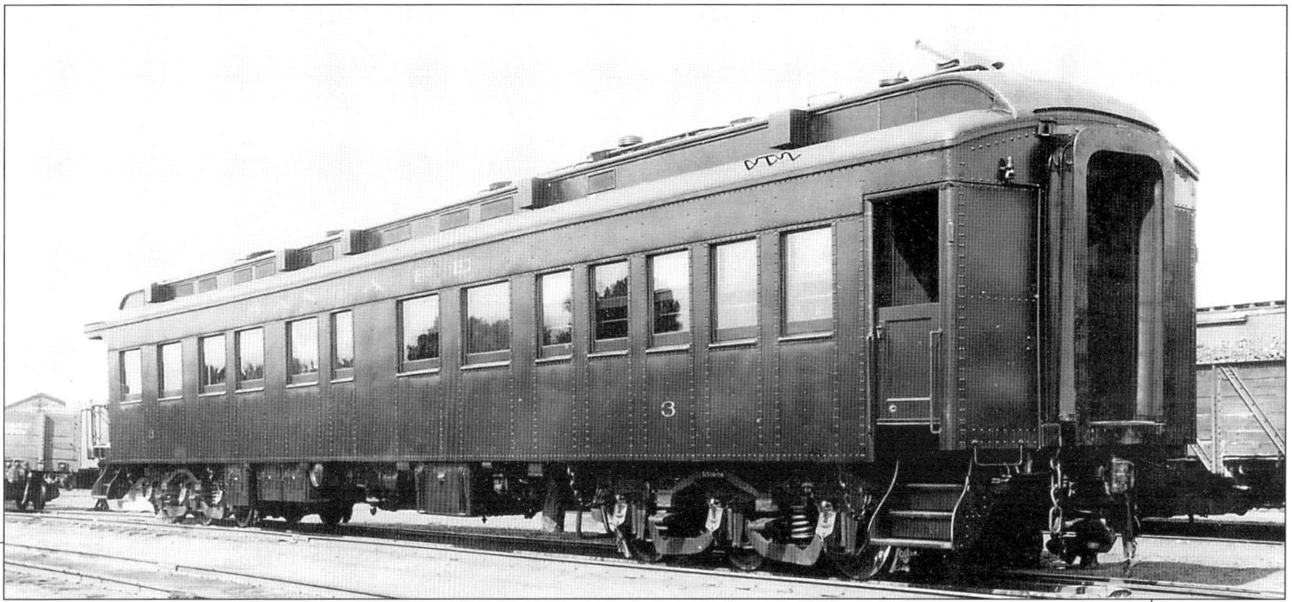

Business car No. 3 was for nearly three decades assigned to the Vice President of Traffic, but was unassigned at the time of this photo, below, taken just prior to it being sold to the Amusement Corporation of America, at Ft. Worth in November 1941. *-John B. McCall collection*

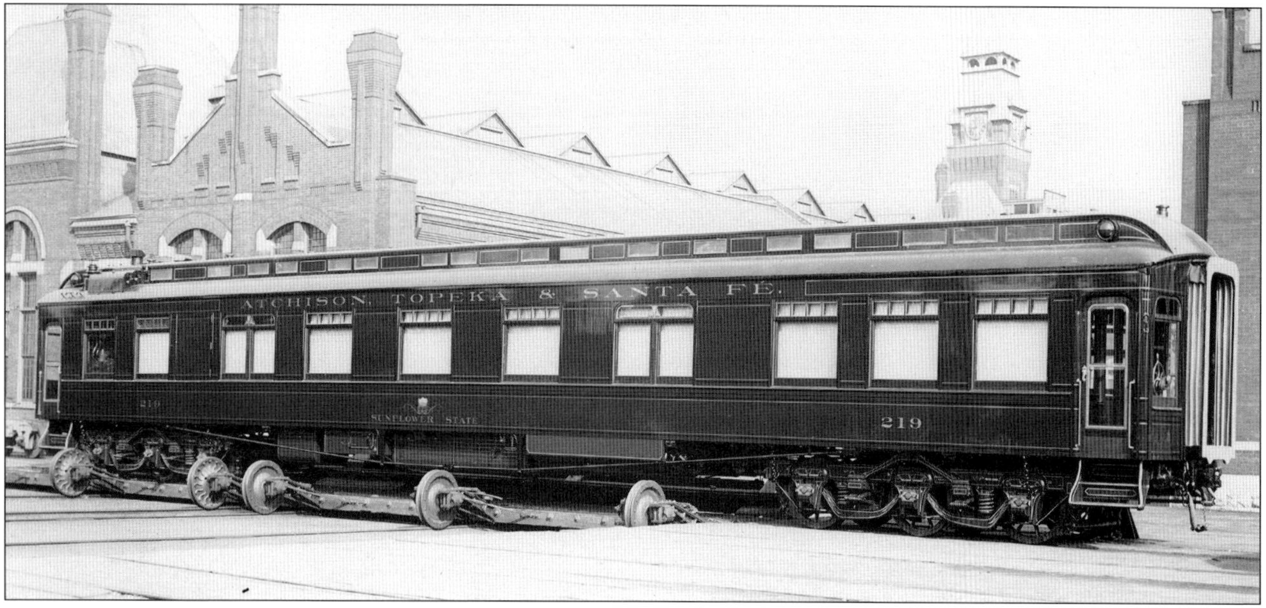

AT&SF Business car 2nd No. 4, started life as 69-foot Business car 219, Sunflower State, built by Pullman for the Santa Fe in 1890. It was constructed for Santa Fe president J. W. Reinhart, whose term was very short-lived. Soon after the Sunflower State and Bay State became part of the property, the old company went into receivership, and Reinhart went along with it. It is interesting to note that two such luxurious cars were purchased when the road was in such difficult financial shape. The six-wheel car was renumbered AT&SF 2 in 1904 and one year later it was renumbered 2nd 4, the number it carried until being converted to a Bunk car in 1952. The car had six single and three double beds with a sleeping capacity of 12 and had a light weight of 137,100 lbs. Notice that Sunflower State has spoked wheels, an item more common on European passenger cars.-*Pullman photo courtesy Smithsonian Institution*

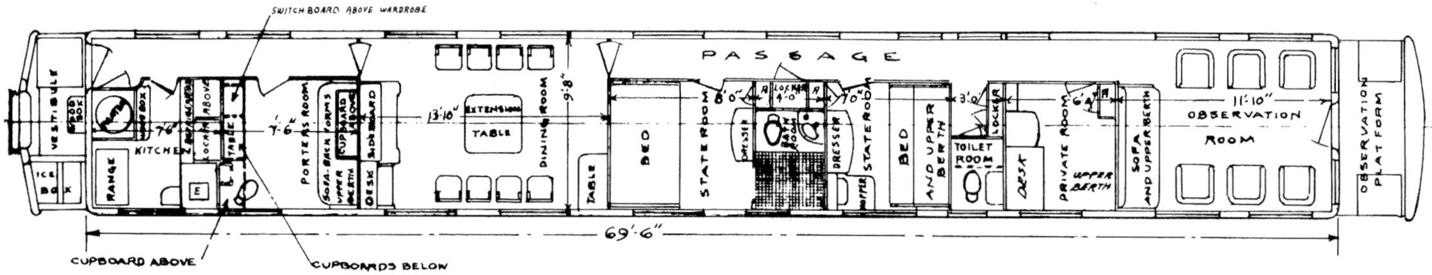

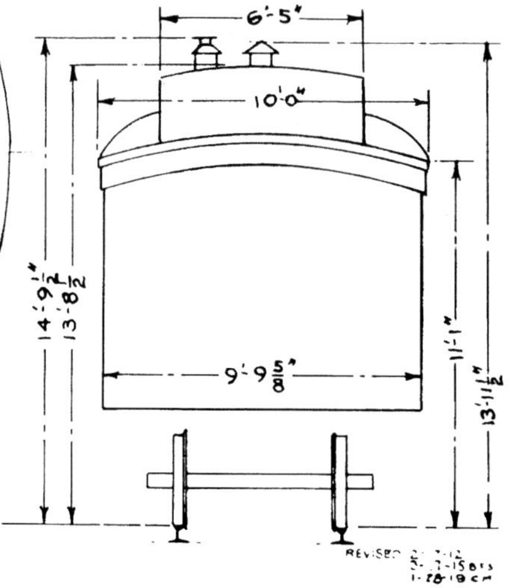

Business Car 2nd No. 4
(former No. 2, No. 219, Sunflower State)
Pullman 1893

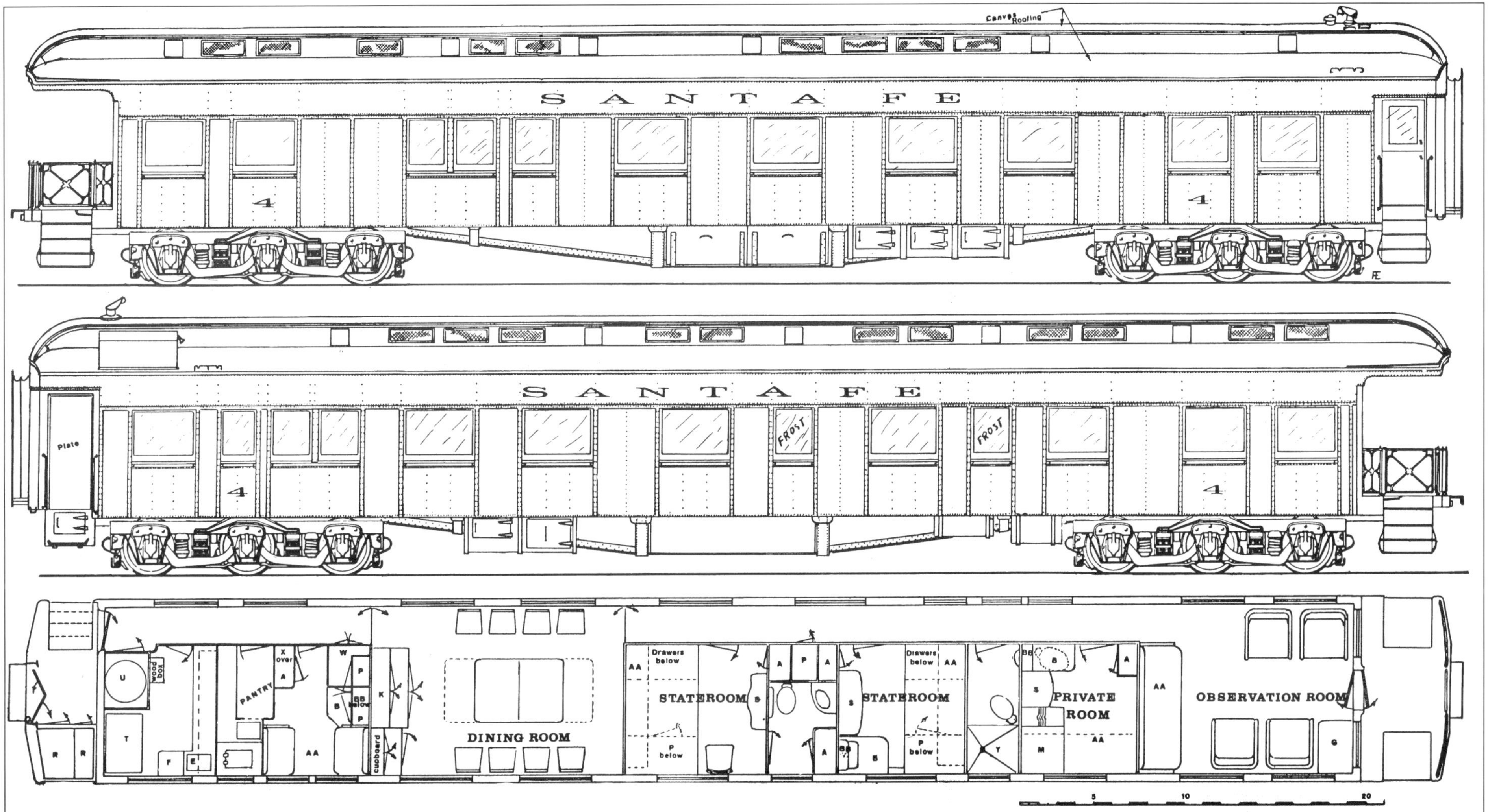

KEY TO DRAWING

A	Wardrobe	I	Chime	R	Refrigerator	Z	Shelves Overhead
B	Hopper	J	Speedometer	S	Dresser	AA	With Upper Berth
C	Washstand	K	Sideboard	T	Range	BB	Folding Washstand
D	Ice Box	L	Annunciator	U	Water Heater	CC	Folding Hopper
E	Sink	M	Secretary's Desk	V	Charcoal Broiler	DD	Radio
F	Drainboard	N	Bookcase	W	Linen/Storage	EE	Convertible Sofa
G	Table	O	Buffet	X	Electric Reg. Locker		
H	Switchboard	P	Locker	Y	Shower		

Business Car 2nd No. 4

As in service as MW 199020

Topeka, Kansas, August 12, 1967

Floorplan data from 23 CS-1622 April 12, 1926

Drawn by Frank M. Ellington

This car, originally constructed by Pullman in 1893 as Business 219 Sunflower State, was renumbered car No. 2 in 1904, then subsequently renumbered (2nd) No. 4 the following year. The car is seen here many years later at Topeka, Kansas, on July 3, 1928. After several major shoppings, the original lines have been almost completely lost. (2nd) No. 4 was assigned to the Vice President of Operations for a number of years, but was assigned to John Purcell, Assistant to Vice President, Mechanical Department when this photo was taken. *-Santa Fe photo, Frank Ellington collection*

Opposite side of (2nd) No. 4 at Topeka, July 3, 1928, after a recent shopping. *-Santa Fe photo, Frank Ellington collection*

Topeka Shops renumbered (2nd) No. 4 to ATSF 199020 in May 1952 to make room for the heavyweight Business car which had just been converted from Coach 3346. The Business car numbered 199020 was then converted to a foreman car 199020 at Cleburne, Texas, in September 1953. After nearly sixty years as a Business car for high Santa Fe officials and almost fifteen years as Foreman car in Maintenance-of-Way service the old "4," nears the end of her career at Topeka, Kansas, below, in August 1967. *-Joseph W. Shine*

Business car No. 5 started life as Barney & Smith Dining car No. 15 (later renumbered ATSF 1408), built in 1891. The 65-foot car was converted to Business car 223 in 1903, and renumbered AT&SF 5, shortly thereafter. The car had eight single and two double beds with a sleeping capacity of 12. It was rebuilt with a 6-inch "I" beam underframe and steel platforms and had a light weight of 127,500 lbs. Notice the believed to be one-of-a-kind observation platform railing. Here Car 5 has emerged from the Topeka Shops in January 1928 resplendent in her new jacket of steel plating and painted passenger green. Number 5 is now ready for many more years of service. The window arrangement remained virtually the same as it was in the days of its wooden sheathing, and the very ornate observation platform railing was retained. At the time the car was assigned to the Assistant General Manager Western Lines, Northern District. *-Santa Fe photos, Frank Ellington collection*

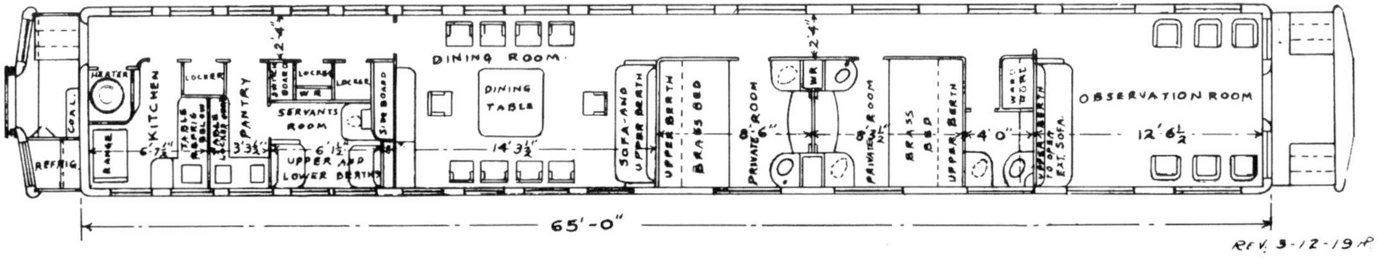

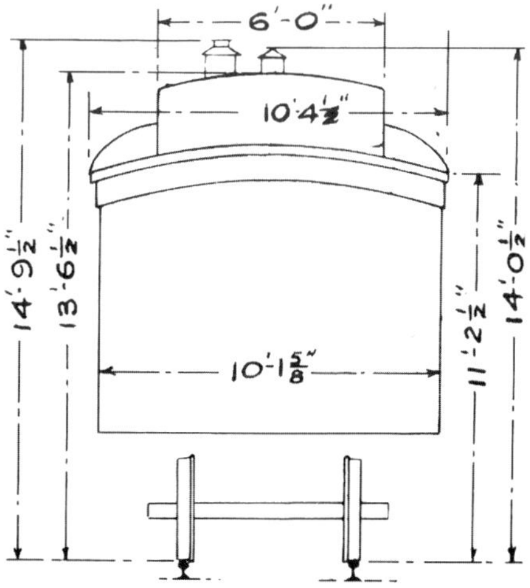

Business Car No. 5
(rebuilt from Diner 1408, No. 223)
Barney & Smith 1891

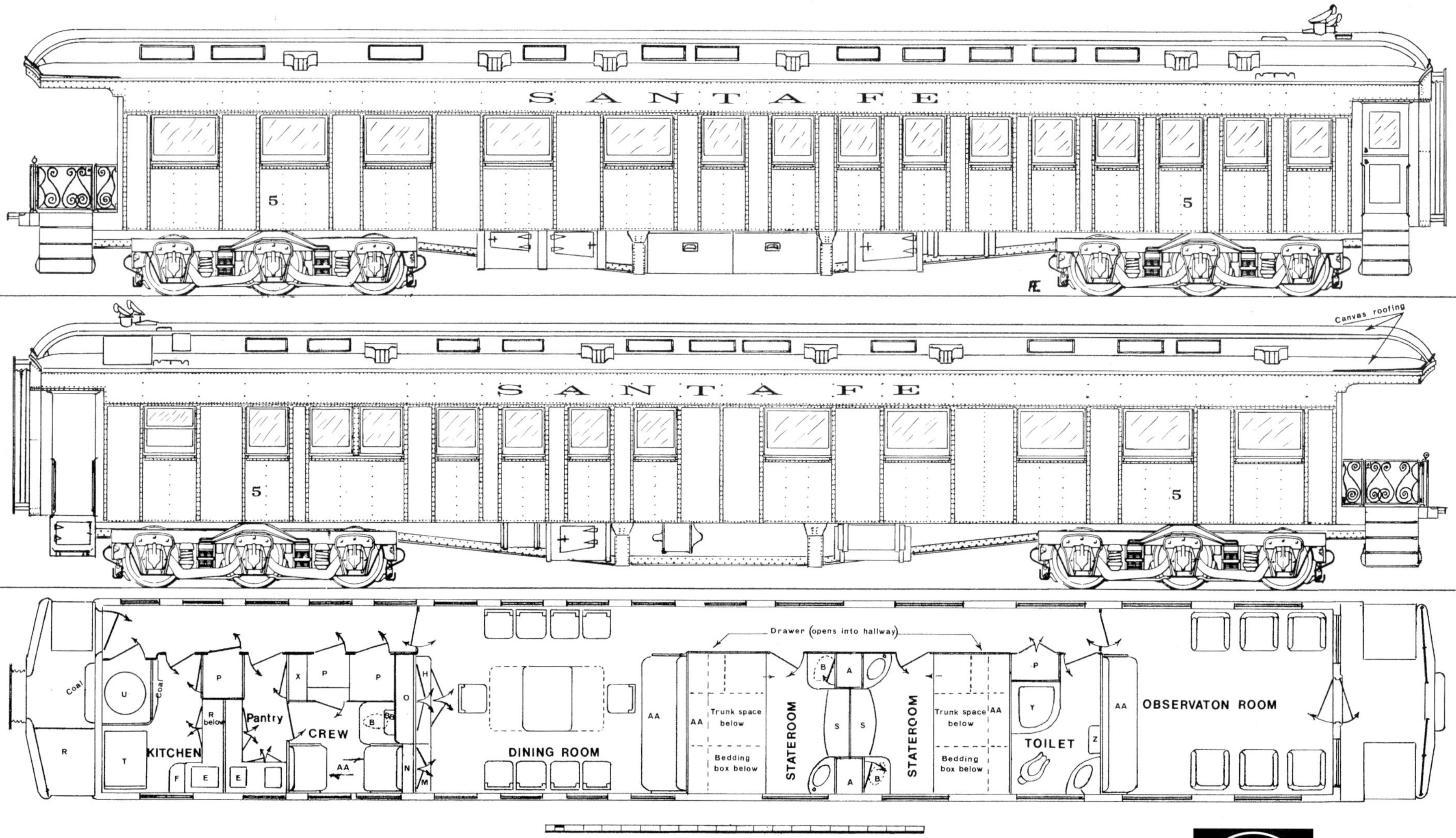

KEY TO DRAWING

A	Wardrobe	I	Chime	R	Refrigerator	Z	Shelves Overhead
B	Hopper	J	Speedometer	S	Dresser	AA	With Upper Berth
C	Washstand	K	Sideboard	T	Range	BB	Folding Washstand
D	Ice Box	L	Annunciator	U	Water Heater	CC	Folding Hopper
E	Sink	M	Secretary's Desk	V	Charcoal Broiler	DD	Radio
F	Drainboard	N	Bookcase	W	Linen/Storage	EE	Convertible Sofa
G	Table	O	Buffet	X	Electric Reg. Locker		
H	Switchboard	P	Locker	Y	Shower		

Business Car No. 5

As in service as MW 199191

Morris, Kansas, September 12, 1968

Floorplan data from 23 C-81 February 16, 1931

Drawn by Frank M. Ellington

While the carbody of the 5 is being made ready inside the Topeka Shops, the steel underframe stands on temporary trucks on the transfer table outside on November 29, 1927. Topeka shops transformed a number of the wood carbody cars into steel-encased cars in the late 1920s. *-Santa Fe photo, Frank Ellington collection*

Can No. 5 was converted to Bunk car 199191 February, 12, 1951. After eighty-one years of service, first as a Dining car, then as a Business car, then as a Maintenance of Way Foreman car, the former No. 5 stands retired at Newton, Kansas, in August 1971, prior to being donated to the Smoky Hill Railway Historical Society in November 1971. *-Joseph W. Shine*

6
A.T.&S.F.

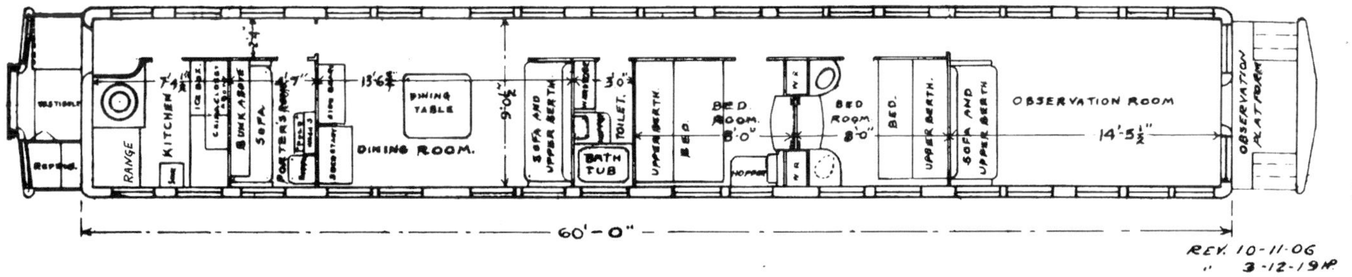

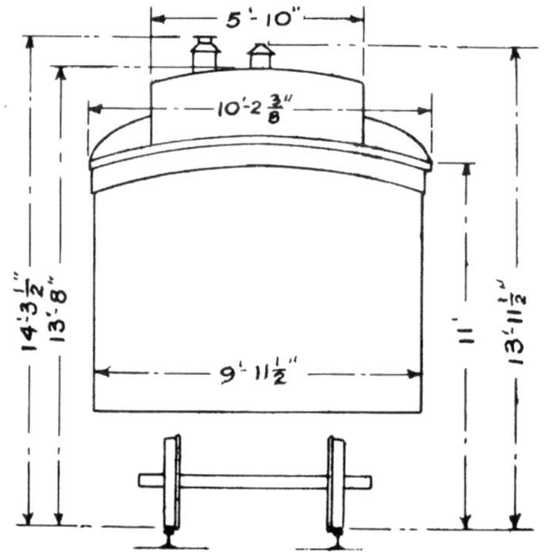

Business Car No. 6
(rebuilt from Coach 101, No. 218)
AT&SF 1887

Santa Fe Business car No. 6, originally AT&SF Coach 101, built by the railroad in 1887, was converted to AT&SF 218, Business car, in June 1894. The 60-foot car was renumbered AT&SF No. 6 in 1903. The car had eight single and two double beds with a sleeping capacity of 12. It was rebuilt with a 6-inch "I" beam underframe and steel platforms and had a light weight of 116,100 lbs. It was converted to 199180 Bunk car in January 1936. No photos have been uncovered of this car.

7
A.T.&S.F.

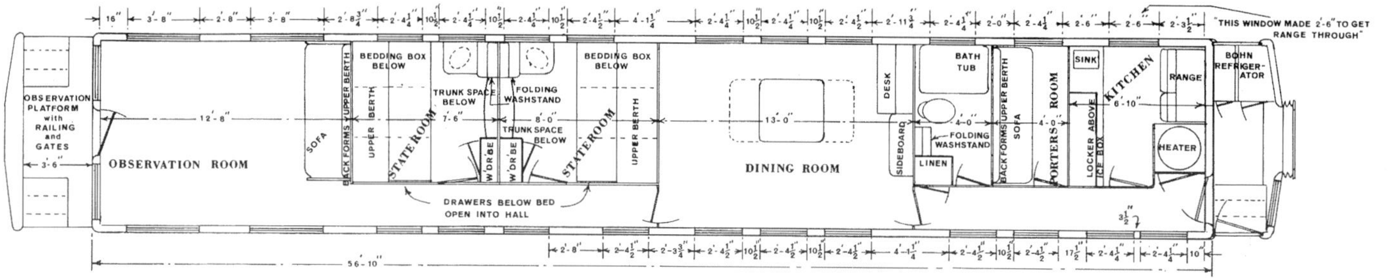

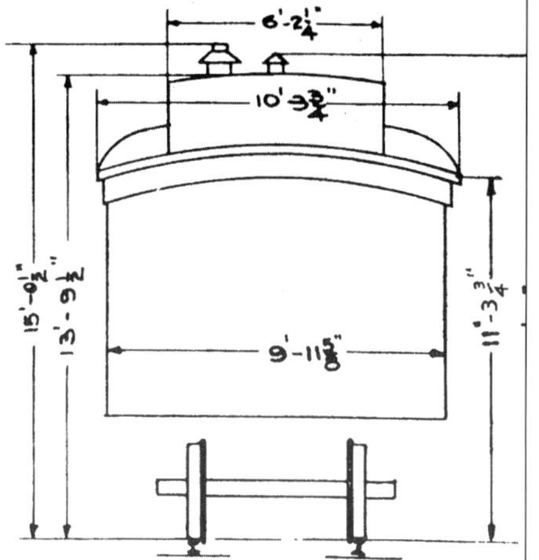

Business Car No. 7
(rebuilt from Coach 99, No. 216)
AT&SF 1882

Santa Fe Business car No. 7, originally AT&SF Coach 99, built by the railroad in 1882, was converted to AT&SF 216, Business car, in June 1894. The 57-foot car was renumbered AT&SF No. 7 in 1905. The car had six single and two double beds with a sleeping capacity of 10. It was rebuilt with a 6-inch "I" beam underframe and steel platforms and had a light weight of 126,300 lbs. It was sold to the GC&SF March 30, 1931, and dismantled in April the following year. No photos have been uncovered of this car.

8
A.T.&S.F.

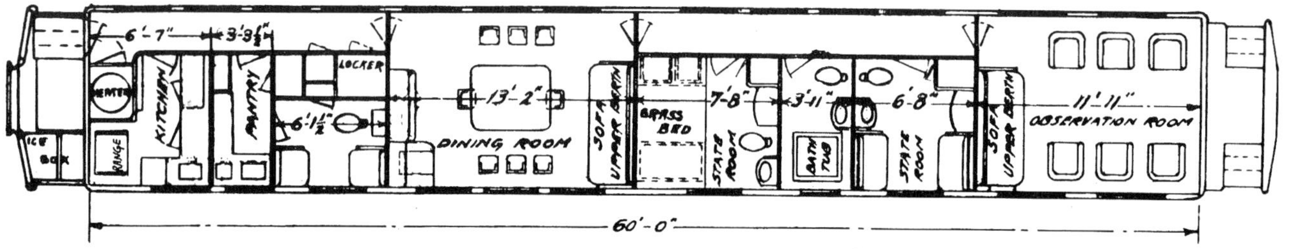

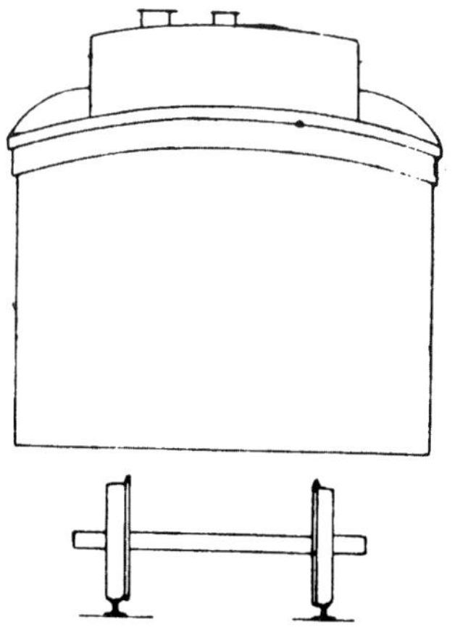

Business Car 2nd No. 8
(rebuilt from Chair 1028)
AT&SF 1899

The first Santa Fe Business car No. 8, originally AT&SF Coach 199, built by the railroad in 1887, was converted to AT&SF 215, Business car, in October 1893. The 51-foot, six wheel car was renumbered AT&SF No. 8 in 1904 and was dismantled March 11, 1918. No photos have been uncovered of this car. Retired in 1918. Car 2nd No. 8, was converted Chair No. 1028 in 1918. This car had eight single and one double beds with a sleeping capacity of 18. It was rebuilt with a 6-inch "I" beam underframe and steel platforms and had a light weight of 127,500 lbs.

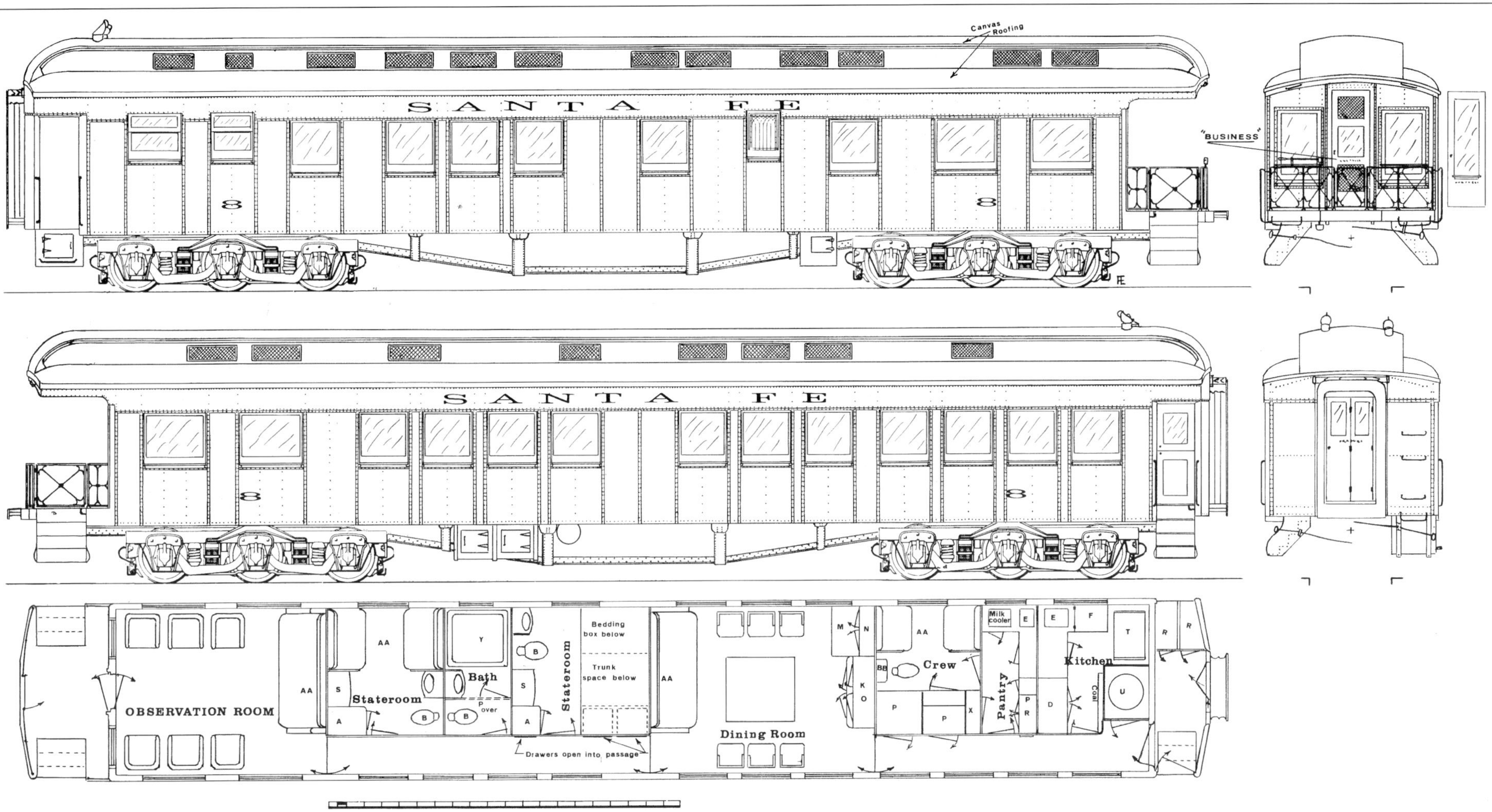

KEY TO DRAWING

A	Wardrobe	I	Chime	R	Refrigerator	Z	Shelves Overhead
B	Hopper	J	Speedometer	S	Dresser	AA	With Upper Berth
C	Washstand	K	Sideboard	T	Range	BB	Folding Washstand
D	Ice Box	L	Annunciator	U	Water Heater	CC	Folding Hopper
E	Sink	M	Secretary's Desk	V	Charcoal Broiler	DD	Radio
F	Drainboard	N	Bookcase	W	Linen/Storage	EE	Convertible Sofa
G	Table	O	Buffet	X	Electric Reg. Locker		
H	Switchboard	P	Locker	Y	Shower		

Business Car 2nd No. 8

As in service as MW 199146
Nortonville, Kansas, June 1969
Floorplan data from 23 C-766 July 31, 1930
Drawn by Frank M. Ellington

Business car 2nd No. 8, which had been converted from Chair car 1028 in July 1918, was again converted, this time to Bunk car 199146 at Cleburne, September 30, 1953. Here we see the right side of old Business "8" February 3, 1962. It was sold for scrap in May 1971. *-John B. McCall*

Santa Fe Maintenance of Way Bunk car 199146 (ex-2nd No. 8) sits on a weed-grown spur in Kansas in the spring of 1971, the last year of company service. Although the car appears to be unchanged on the exterior sides, the addition of smoke jacks midway and in the observation lounge area indicate that additional stoves have been added and the interior modified significantly. The car was stenciled "Rear end movement only." The TV antenna adds a nice touch. *-Three photos, Bill Childers*

9
A.T.&S.F.

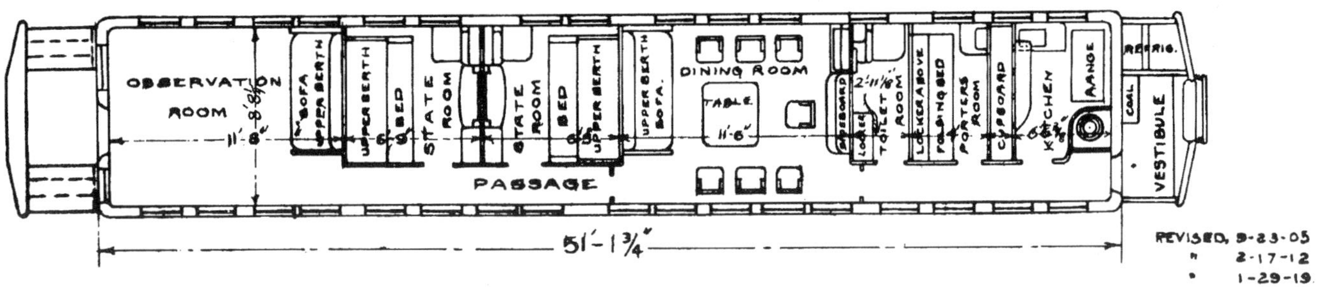

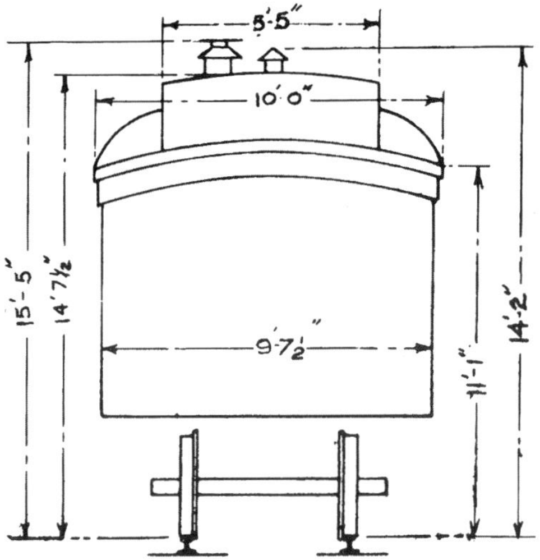

Business Car No. 9
(rebuilt from, A&P 99, SFP 99)
H&H 1882

Santa Fe Business car No. 9, originally Atlantic & Pacific 99, built by H&H in 1882, was renumbered to Santa Fe Pacific 99, Business car, in June 1897. The 51-foot car was renumbered AT&SF No. 9 in July 1902. The car had seven single and two double beds with a sleeping capacity of 11. It was rebuilt with a 6-inch "I" beam underframe and steel platforms and had a light weight of 131,200 lbs. It was renumbered to Santa Fe Superintendent's car 2nd 412 in April 1927. No photos have been uncovered of this car.

10
A.T.&S.F.

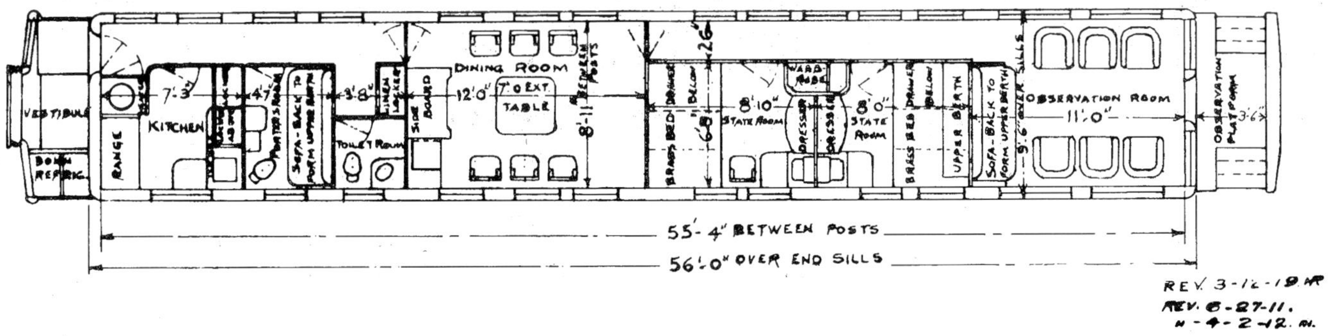

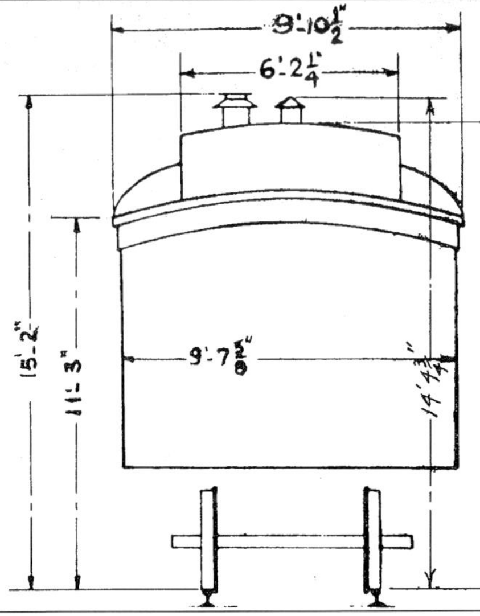

Business Car No. 10
(originally A&P 100, SFP 100)
A&P 1883

Santa Fe Business car No. 10 was originally Atlantic & Pacific Business car 100 constructed by the Atlantic & Pacific Railroad in 1883. The 56-foot car was re-equipped with six-wheel trucks, and became AT&SF Business car No.10 in July 1902, when it was assigned to the Superintendent of Transportation. The car had five single and two double beds with a sleeping capacity of 9. It was rebuilt with a 6-inch "I" beam underframe and steel platforms and had a light weight of 123,600 lbs. Later it was assigned to the General Manager, Santa Fe Prescott & Pacific. During the 1920s it was assigned to the Division Superintendent of the Arizona Division and it was renumbered to Santa Fe Superintendent's car 2nd 413 in September 1927. The car is seen here in service as A&P 100, circa-1895, prior to being equipped with six wheel trucks. *-Library State Historical Society of Colorado, Vernon Glover collection*

11
G.C.&S.F.

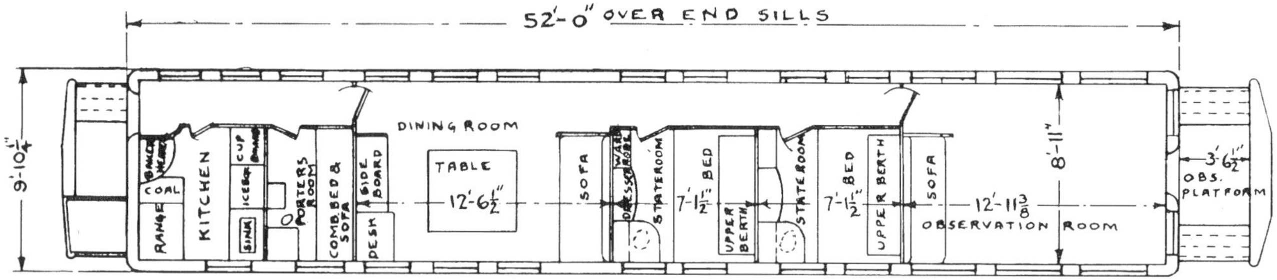

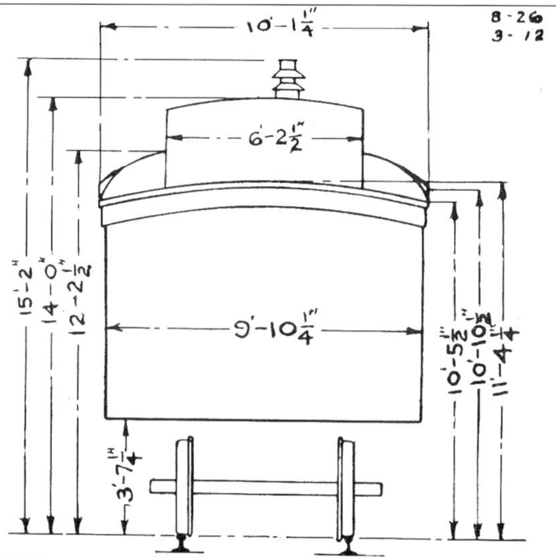

GC&SF Business Car 2nd No. 11
(originally, GC&SF 102, GC&SF 13)
B&S 1883

The first GC&SF Business car 11 was built by Barney & Smith in 1890, but was dropped from the roster in 1905. Santa Fe Business car 2nd No. 11, originally Business car GC&SF 102, built by Barney & Smith in 1883. The 50-foot car was renumbered GC&SF No. 13 in 1904 and then GC&SF 2nd No. 11 in 1905. The car had seven single and two double beds with a sleeping capacity of 11. It was rebuilt with a 6-inch "I" beam underframe and steel platforms and had a light weight of 121,500 lbs. It was renumbered to ATSF 11 on May 27, 1931 and dismantled that December. No photos have been uncovered of this car.

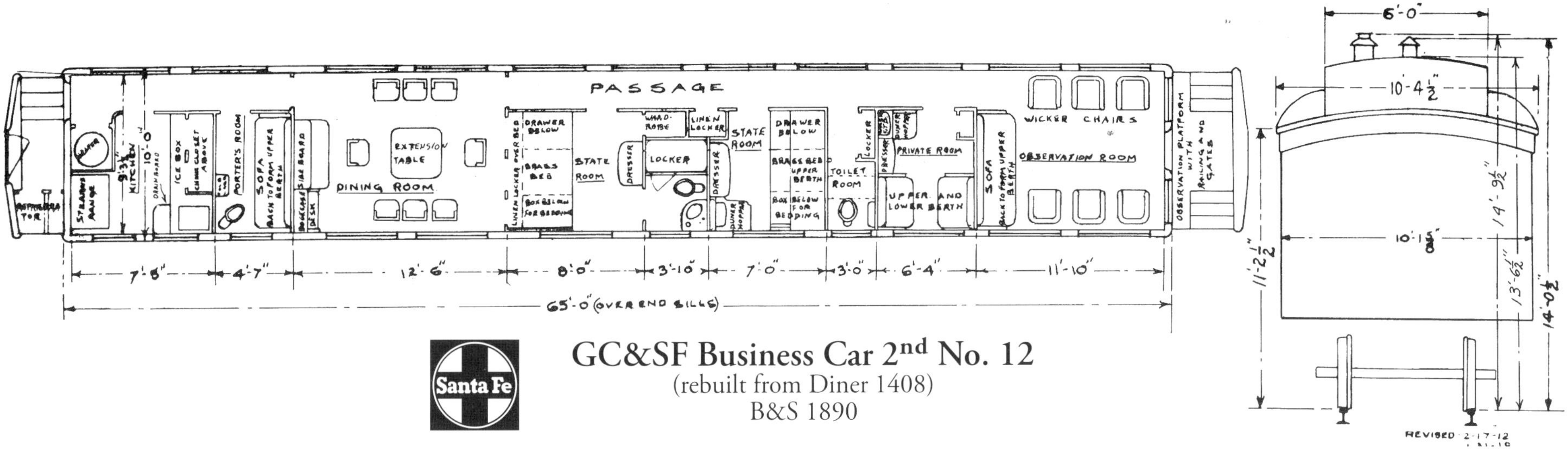

GC&SF Business Car 2nd No. 12

(rebuilt from Diner 1408)
B&S 1890

The first GC&SF Business car 12 was built by Barney & Smith in 1886, was originally AT&SF Business car 100, renumbered 217 in January 1895, and GC&SF 12 in 1903. It was again renumbered ATSF 20 in 1920. Santa Fe Business car 2nd No. 12, at Topeka in 1950, was converted in 1910 from Diner 1408, built by Barney & Smith in 1890. The 65-foot car had six single and three double beds with a sleeping capacity of 12. It was rebuilt with a 6-inch "I" beam underframe and steel platforms and had a light weight of 139,400 lbs. It was renumbered to ATSF 12 on May 27, 1931. *-Ray Hilner*

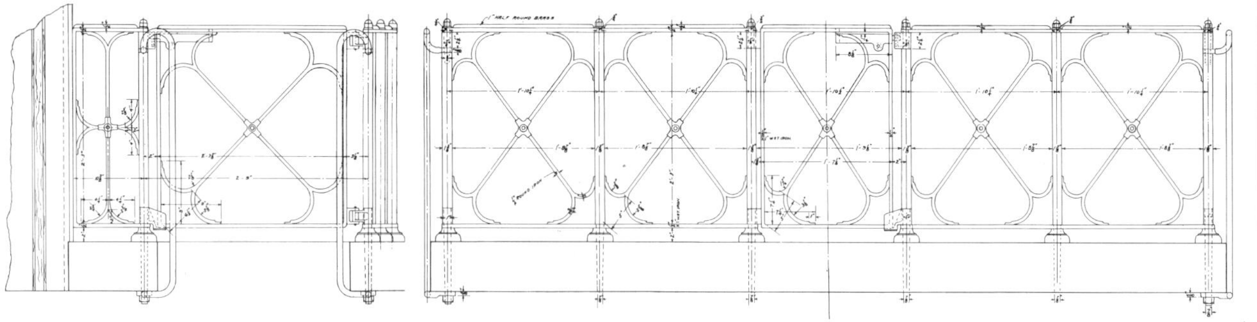

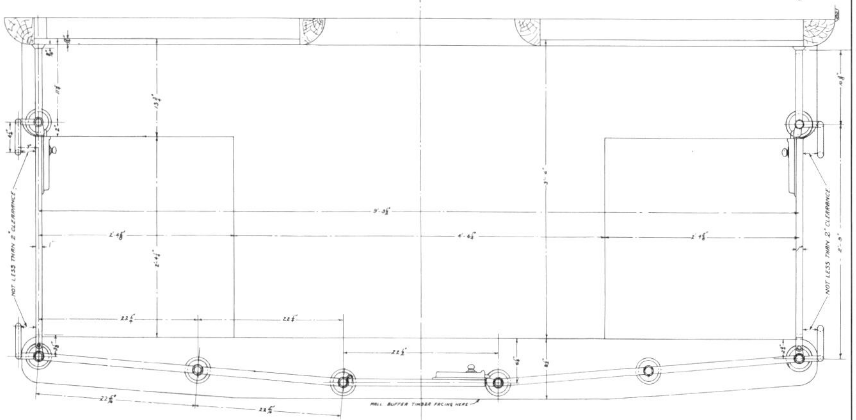

Santa Fe

Railing and Gates for Business Car No. 12

Data from 23 A-363 April 9, 1912

Drawn by Frank M. Ellington

Steel-plated Business car No. 12 at Topeka prior to being converted to Shovel Operator's car 199199 in February 1953.
-Ray Hilner

KEY TO DRAWING

A	Wardrobe	I	Chime	R	Refrigerator	Z	Shelves Overhead
B	Hopper	J	Speedometer	S	Dresser	AA	With Upper Berth
C	Washstand	K	Sideboard	T	Range	BB	Folding Washstand
D	Ice Box	L	Annunciator	U	Water Heater	CC	Folding Hopper
E	Sink	M	Secretary's Desk	V	Charcoal Broiler	DD	Radio
F	Drainboard	N	Bookcase	W	Linen/Storage	EE	Convertible Sofa
G	Table	O	Buffet	X	Electric Reg. Locker		
H	Switchboard	P	Locker	Y	Shower		

Business Car 2nd No. 12

As in service as MW 199199

Newton, Kansas, June 1969

Floorplan data from 23 C-1693 July 31, 1930

Drawn by Frank M. Ellington

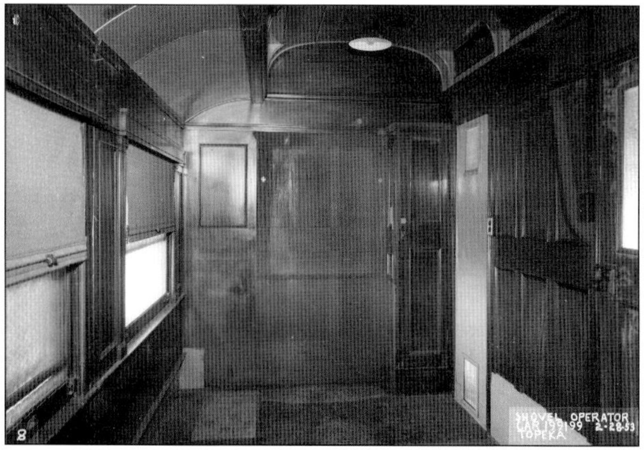

Car 2nd No. 12 was converted to Shovel Operator's car 199199 at Topeka in February 1953. The wood trim remains, but its interior furnishings were transferred to one of the heavyweight coach rebuilds. Steam and signal lines have been removed necessitating the installation of several oil heaters. The car was eventually sold for scrap in April 1972. *-All, John B. McCall collection*

13
G.C.&S.F.

GC&SF 13, originally built as GC&SF 102 General Manager's car by Barney & Smith in 1883 was renumbered GC&SF 2nd No. 11 in 1905. See car No. 11.

14
A.T.&S.F.

Santa Fe Business car No. 14, originally Southern California Railway Officer's car 98 built by Pullman in 1881, was renumbered Santa Fe, Prescott & Phoenix No. 14 in 1904 and renumbered 14 in 1912. The 60-foot car was dismantled March 7, 1919. No photos have been uncovered of this car.

15
A.T.&S.F.

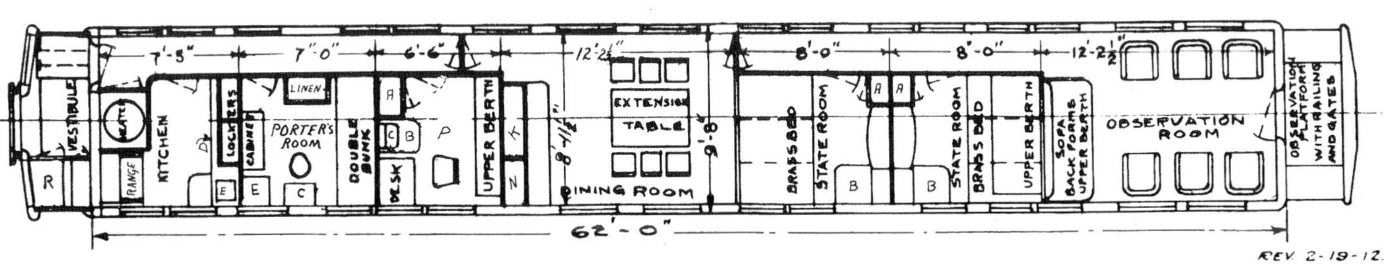

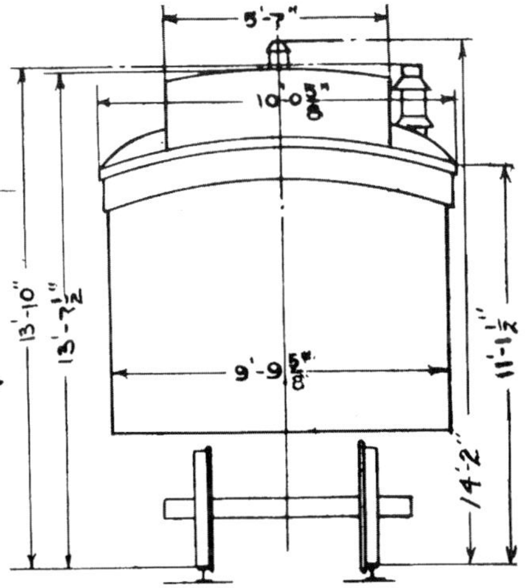

Business Car No. 15
(originally E Ry of NM 15)
Pullman 1907

Company records show that Santa Fe Business car No. 15, originally Eastern Railway of New Mexico Officer's car 15, was built by Pullman in 1907, and was renumbered to Santa Fe 15, in January 1912. The 62-foot car had six single and two double beds with a sleeping capacity of 10. It had a light weight of 125,100 lbs. The first assignment of the "15" back in 1907 was to the President and General Manager of the Pecos Lines, which consisted of the line north of Pecos to the New Mexico/Texas state line and east to Amarillo. The following year it served the Superintendent of the Eastern Ry. of New Mexico which had just been completed between Clovis and Belen, New Mexico. It is seen here at San Bernardino, California, on March 16, 1945.
-Jack Whitmeyer photo, SFRH&MS Collection

The Mystery of No. 15

Herewith are two company photos of Business car No. 15, taken at Topeka October 31, 1927, after the old wooden car had received steel veneer plating, cast steel under frame and newer trucks. Note that the 15 had managed to retain, at this point, the older two-piece window sash arrangement as in the wood side days. No. 15 was the only representative of the steel plated equipment in business service that could boast of this window style. All of the former Barney & Smith Diners rebuilt into Business cars that survived the 'twenties got the steel plate veneer (but none had any transom windows exposed afterwards), cast steel under frames and newer trucks. *-Two photos, John B. McCall collection*

There is more to No. 15 than its exposed transom windows however. Santa Fe records show it as being built by Pullman in 1907, but it is appears to have been a much older car than that. It is subtly different from the Business cars that were rebuilt right after 1900 from the early Barney & Smith 1400 series Dining cars. Not only is it different, but there is not the slightest suggestion its roots could be traced to the Dining car roster—and none of the Diners are unaccounted for. Was the 15 built by Pullman in 1907, as Santa Fe would have us believe? Or was it rebuilt from an earlier car. One possible clue is that it was assigned to the Eastern Railway of New Mexico as their primary car at the time.

It has been suggested that it may have come from James Hagerman's car, Pecos Valley and North Eastern No. 77, Hesperia. The window pattern in quite similar to the Hesperia (although that car had open platforms at both ends at the time of its construction). And there is no record of what happened to the Hesperia. No. 15 may, however, have been built from the remains of a wrecked wood Diner, salvaged but listed as destroyed on the books. Yes, the No. 15 is a mystery car.

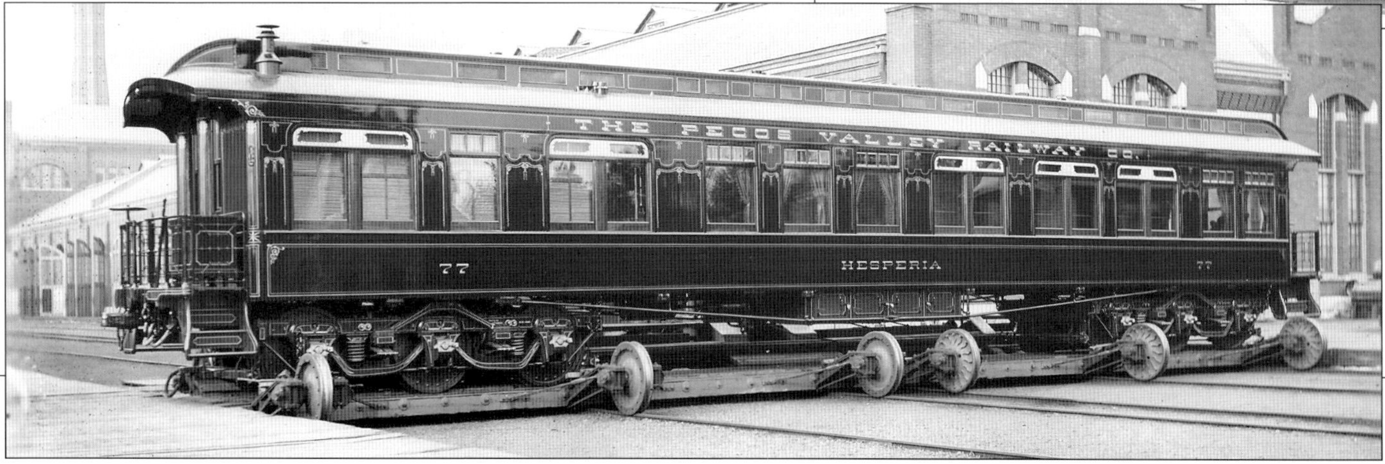

James Hagerman, promoter of the Pecos Valley RR. in 1890, purchased this private car from Pullman in 1892. Hagerman, aided by the Santa Fe, in 1895 built north to meet the Santa Fe at Texico. History does not recall what happened to the car Hesperia leading to the speculation about No. 15. *-Pullman photo courtesy Smithsonian Institution*

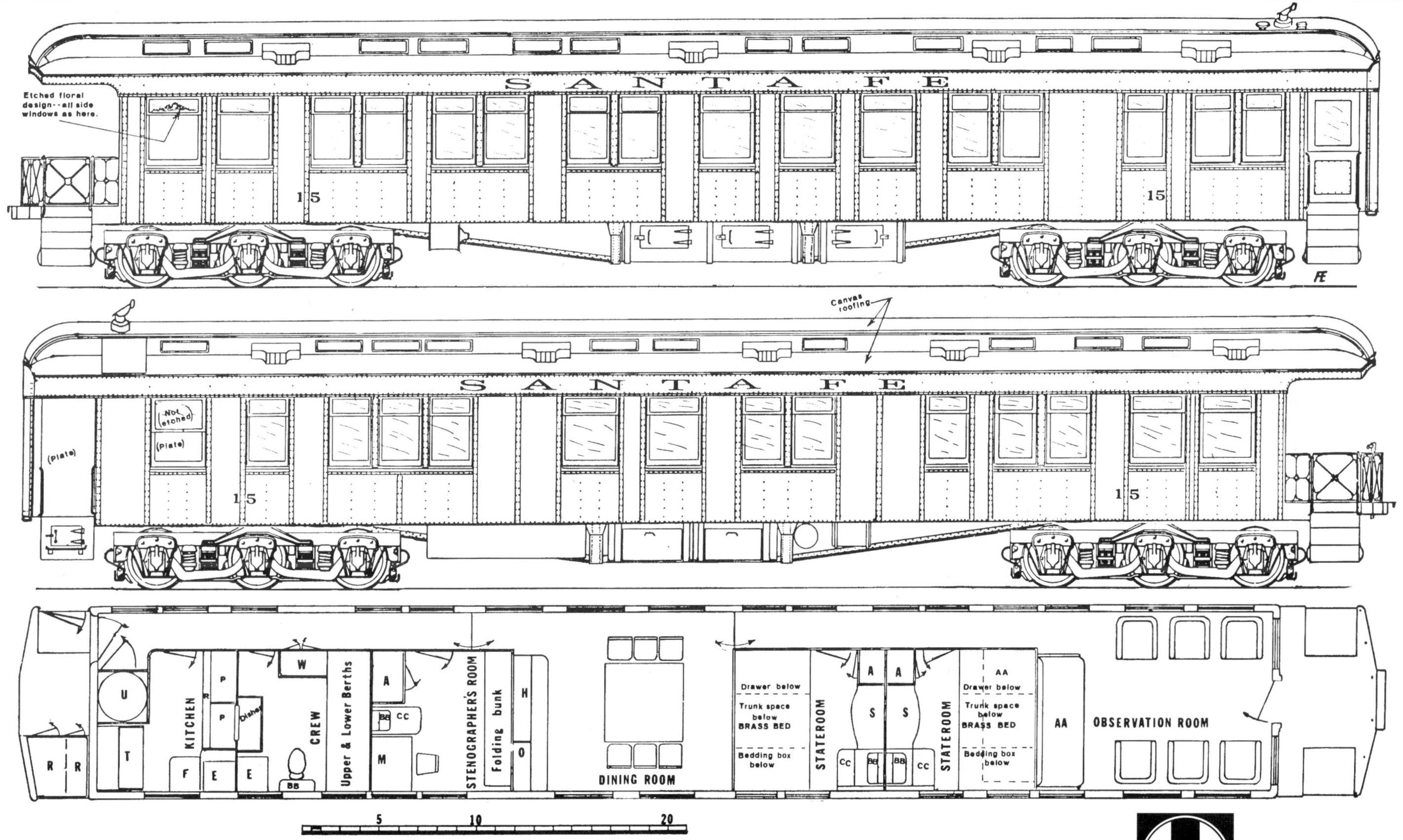

KEY TO DRAWING

A	Wardrobe	I	Chime	R	Refrigerator	Z	Shelves Overhead
B	Hopper	J	Speedometer	S	Dresser	AA	With Upper Berth
C	Washstand	K	Sideboard	T	Range	BB	Folding Washstand
D	Ice Box	L	Annunciator	U	Water Heater	CC	Folding Hopper
E	Sink	M	Secretary's Desk	V	Charcoal Broiler	DD	Radio
F	Drainboard	N	Bookcase	W	Linen/Storage	EE	Convertible Sofa
G	Table	O	Buffet	X	Electric Reg. Locker		
H	Switchboard	P	Locker	Y	Shower		

Business Car No. 15

As in service as MW 199148 (except with upper sash)
Abeline, Kansas, August 1962
Floorplan data from 23 C-695 Rev. C June 1, 1927
Drawn by Frank M. Ellington

A subsequent change which altered the appearance of No. 15 considerably was the addition of a wide steel plate the length of the car above the windows covering the transom windows. Topeka, Kansas, October 1950.
-F.O. Kelley collection

During the 1950s and 1960s most of the Maintenance of Way cars were converted at Cleburne, Texas, but the 15 was converted to a living car for a Shovel Engineer on eastern lines at the Topeka shops in 1953. Here the former No. 15 poses for its picture as ATSF 199148 on August 31, 1953. The car served nearly 20 more years and was retired and sold for scrap in February 1972.
-Santa Fe photos, John B. McCall collection

Interior photos of former Business car No. 15, stripped of furnishings, steam and signal lines, and fitted with oil heaters in the observation lounge and dining room, is seen at Topeka on August 20, 1953. From upper left, the original woodwork in the side board area of the dining room has been retained, and the hall, right, appears much the same, except for the heater and stove pipe in the distance located in the observation lounge. The next frame is of the heater at the rear of the car, and at left, the heater which occupied considerable space in the dining room. Note the fire-resistant metal sheeting surrounding the stove installation. *-Four photos, Santa Fe, John B. McCall collection*

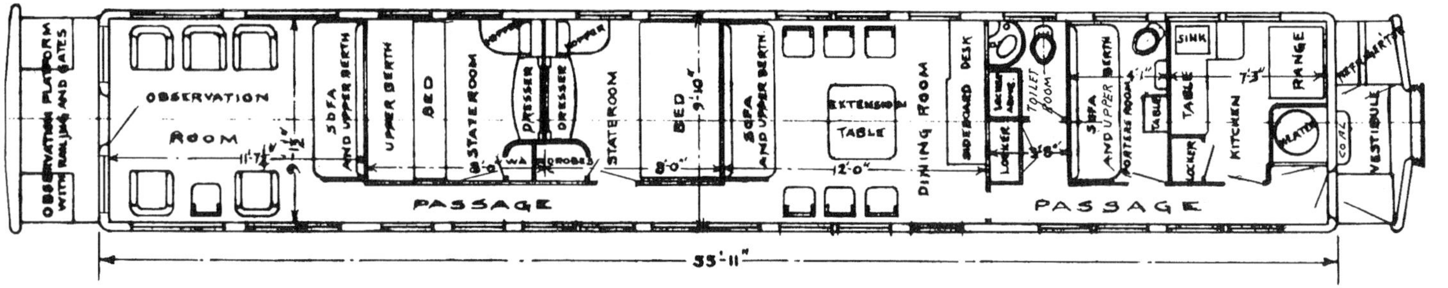

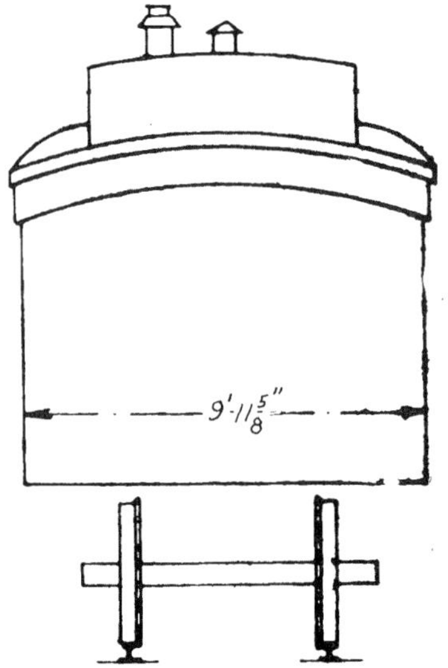

Business Car 1st No. 16
(rebuilt from Coach 2697)
B&S 1890

REV. 2-17-12
" 9-12-19 HP.

Business car 1st No. 16 was originally ATSF Coach 2697, built by St. Louis Car Co. in 1886, and was converted to Business car No. 16 in 1907. The car had seven single and two double beds with a sleeping capacity of 11 and had a light weight of 120,000 lbs. It was dismantled August 28, 1937. The car is seen near the end of its career. *-John B. McCall collection*

17
A.T.&S.F.

One of the Pullman Company's most elegant private cars was the 17 built for Santa Fe President Edward Payson Ripley in 1907 on plan 2248. The 72-foot car was painted "Pullman Green," with a black roof. All striping, lettering and numbers were gold and side grab irons and observation railing polished brass. It served as the President's car until 1918 when the first all steel heavyweight business cars arrived. *-Pullman photo, Frank Ellington collection*

At far left is view of President Ripely's room-filling brass double bed in stateroom "B" at the time of its release from Pullman on April 19, 1907. Car No. 17 had five single berths and one double bed and could sleep seven. *-Pullman photo, Frank Ellington collection*

At left are the more modest facilities in private room "C" looking toward the rear of the car. During the day one used the Davenport lounge and a single berth dropped down for sleeping at night. The cabinet in the center concealed a folding washstand, while the settee at lower right concealed the hopper. *-Pullman photo, Frank Ellington collection*

In this view looking toward the front of the car, can be seen the dining room on the 17 as it was outfitted at Pullman. Against the bulkhead is the highly-polished mahogany side board, china and silver cabinet, and at left, the secretary's desk. The galley and crew quarters are to the right through the passage. Note the intricate stained glass work in the upper windows.

Looking toward the rear of the car, the passage at right leads to the observation lounge. The sofa against the bulkhead could be raised to form an upper bunk. The dining room table, center, could be extended to accommodate as many as eight people sitting in the rather ornate chairs upholstered in leather visible in the picture. *-Two photos, Pullman, Frank Ellington collection*

These two views of the observation lounge at the rear of Business car No. 17 show the car as it was delivered in April 1907. At left, looking toward the unusually spacious rear observation platform, the general arrangement of furnishings which included seven wicker chairs, and a speedometer and air gauge mounted to the left of the rear door. The ceiling fixture contained both gas and electric lights. At right is the same room looking toward the front of the car. Furnishings included a small table and a sofa across the bulkhead which contained an upper berth. The hallway at left led past rooms B and C to the dining room, which can be seen in the distance. -*Two photos, Pullman, Frank Ellington collection*

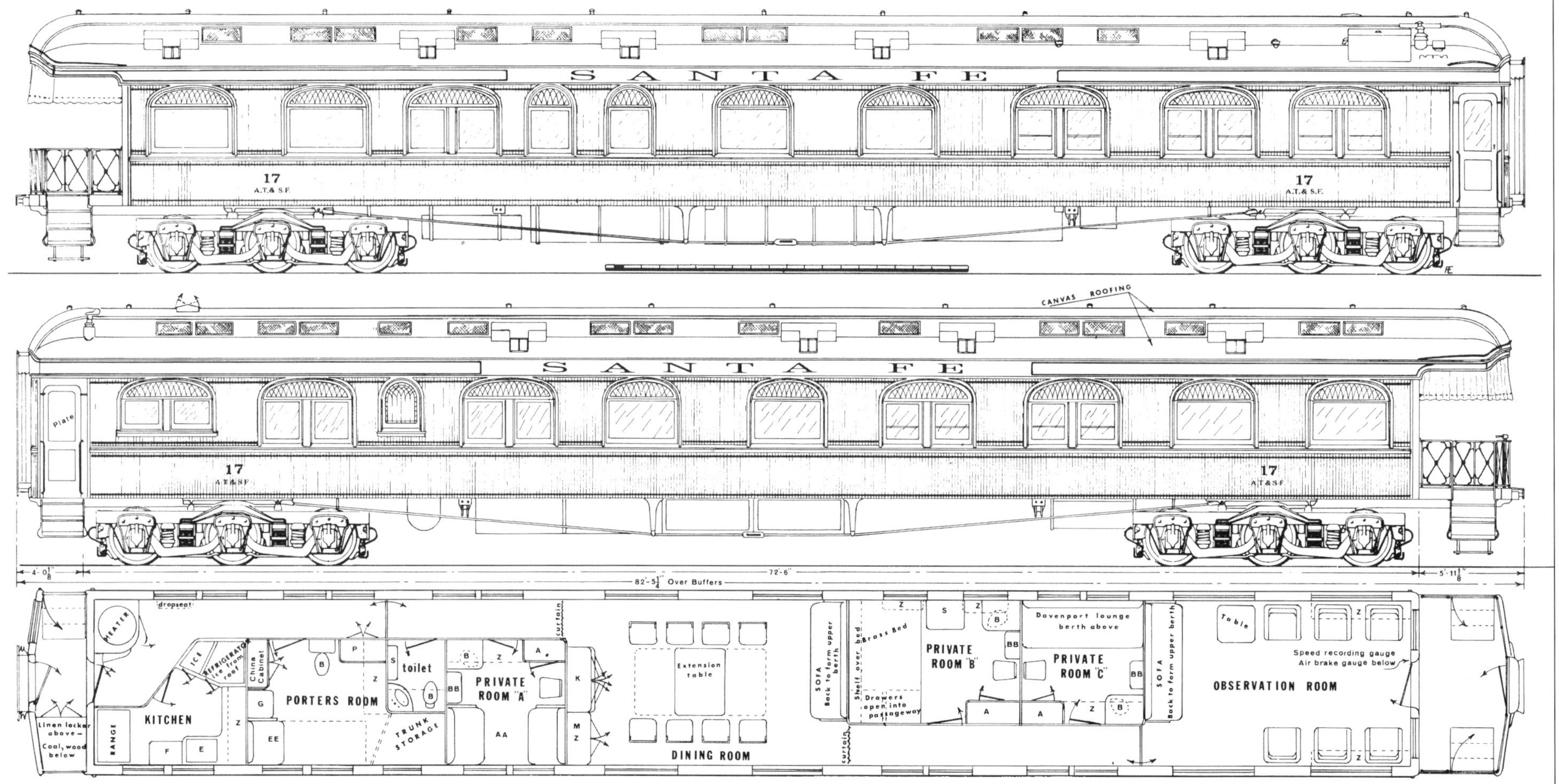

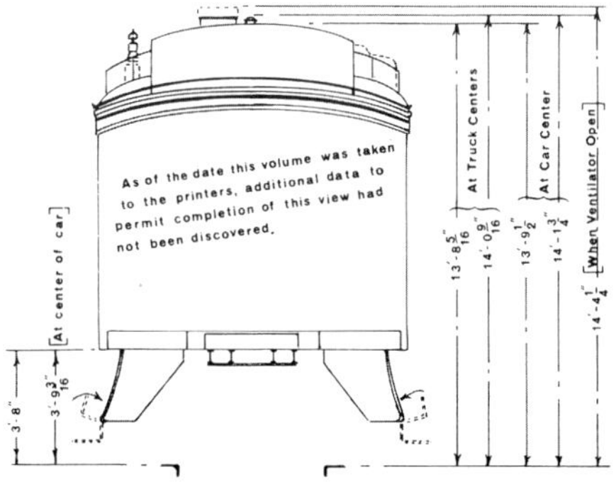

KEY TO DRAWING

A	Wardrobe	K	Sideboard	V	Charcoal Broiler
B	Hopper	L	Annunciator	W	Linen/Storage
C	Washstand	M	Secretary's Desk	X	Electric Reg. Locker
D	Ice Box	N	Bookcase	Y	Shower
E	Sink	O	Buffet	Z	Shelves Overhead
F	Drainboard	P	Locker	AA	With Upper Berth
G	Table	R	Refrigerator	BB	Folding Washstand
H	Switchboard	S	Dresser	CC	Folding Hopper
I	Chime	T	Range	DD	Radio
J	Speedometer	U	Water Heater	EE	Convertible Sofa

Business Car No. 17

Pullman Plan 2248 February 25, 1907

Drawn by Frank M. Ellington

The Porter's Room, facing toward the galley, upper left, was as well appointed as the rest of the car. To the left above is an upper berth, folded into the wall out of the way for daytime travel. At the opposite end of the Porters Room, right, was a large recess for trunk and luggage storage protected by brass rods. The galley contained facilities for preparation of quite elaborate meals. At left we see the coal-fired range with the warmer and oven. At right is the galley, with drain board, sink, and ice refrigerator. *-Four photos, Pullman, Frank Ellington collection*

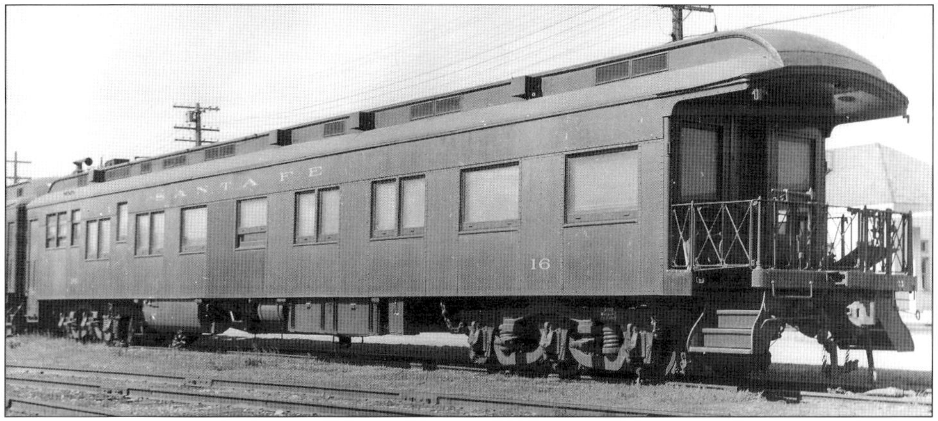

Car 17 was renumbered 2nd 16 in November 1952 to make way for a new heavyweight No. 17 coming on line at that time. The car, as 2nd 16, is seen, left, at Topeka in March 1955. Note the plating now covering the arched transom windows, as well as the extended observation platform *-F.O. Kelley, John B. McCall collection*

Car 2nd No. 16 (ex-17) was converted to Bunk car 194020 at Cleburne in June 1960. F.O. Kelley captured it on film at an unknown location prior to being destroyed by fire in June 1964. *-John B. McCall collection*

Santa Fe Business car No. 18, originally built by Barney & Smith in 1890, was converted from Diner 1404 in 1907. The 65-foot car had eight single and two double beds with a sleeping capacity of 12. It was rebuilt with a 6-inch "I" beam underframe and steel platforms and had a light weight of 145,300 lbs. Car 18 is seen in service on the rear of the Second District Local at San Bernardino on January 18, 1949.
-Jack Whitmeyer photo, SFRH&MS Collection

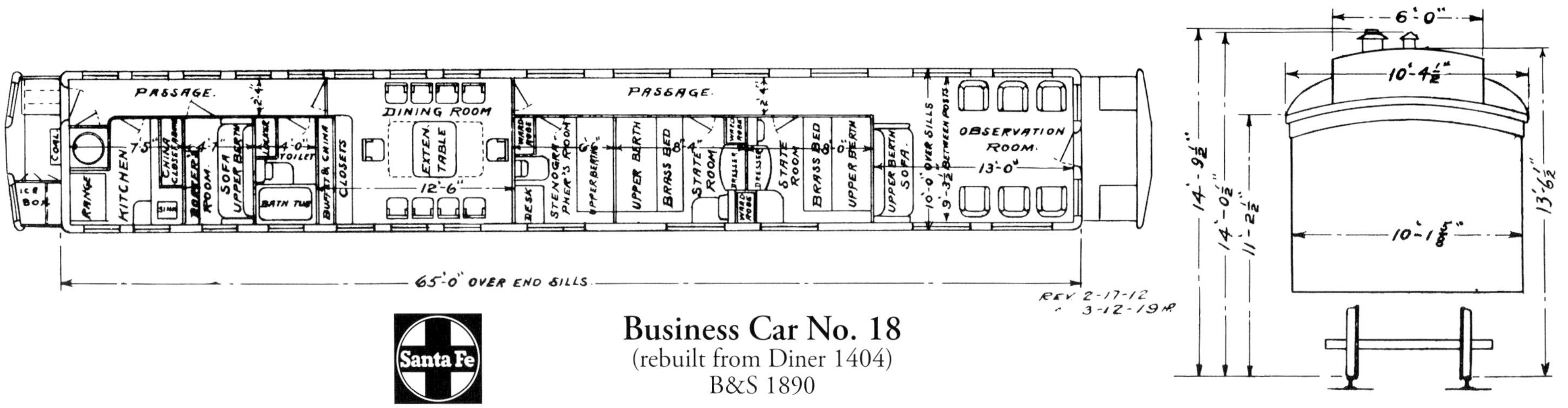

Business Car No. 18
(rebuilt from Diner 1404)
B&S 1890

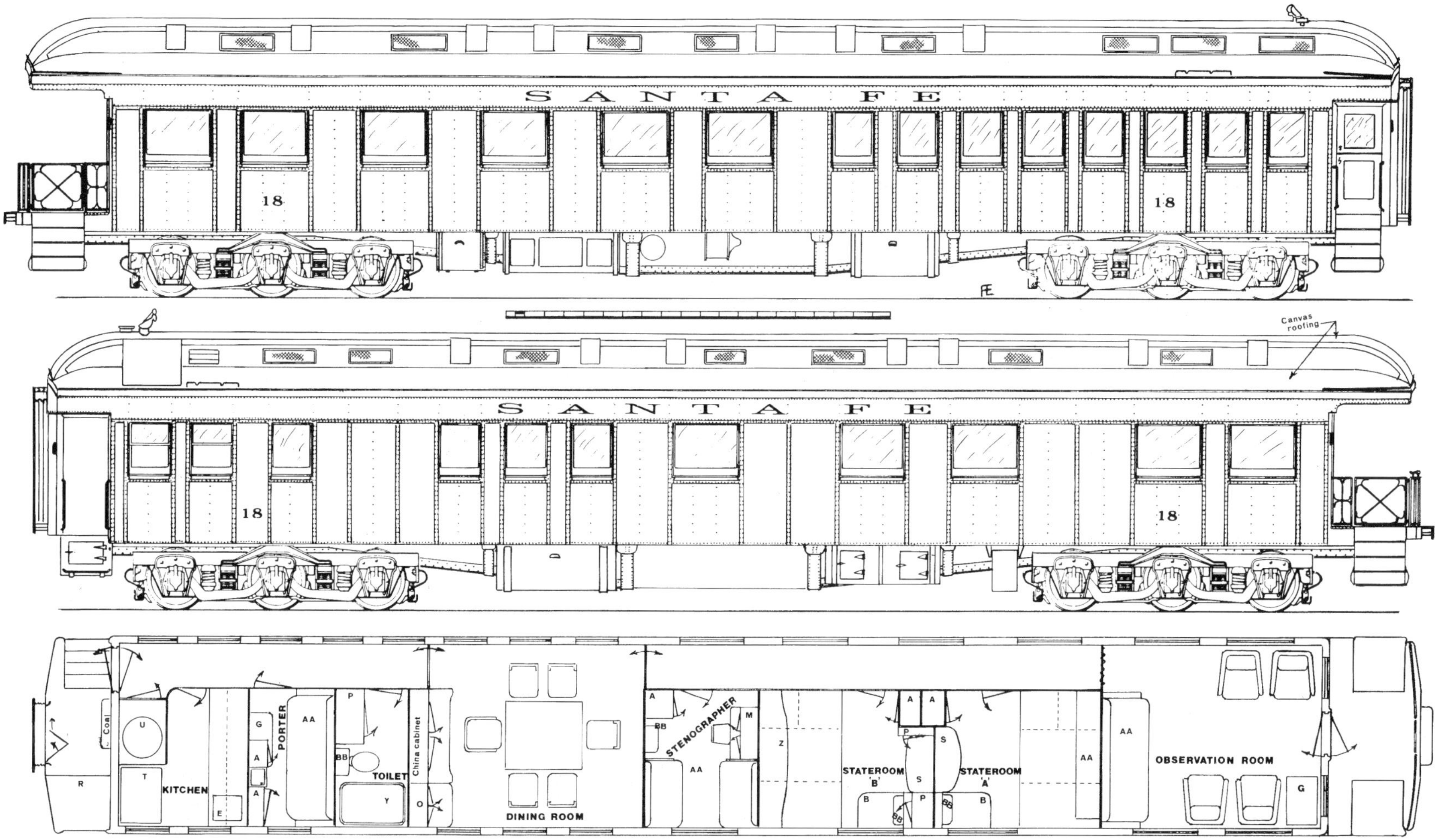

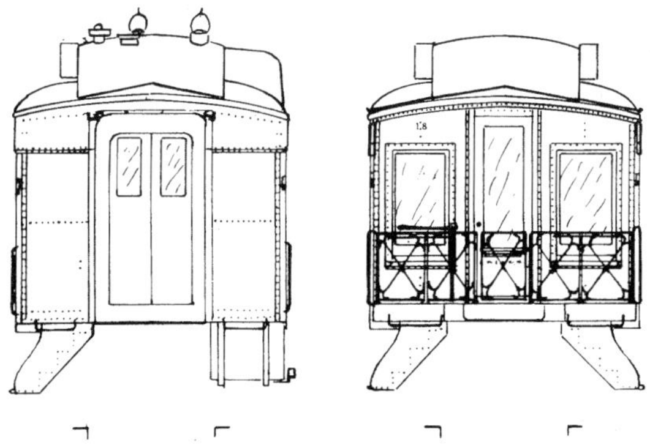

KEY TO DRAWING

A	Wardrobe	K	Sideboard	V	Charcoal Broiler
B	Hopper	L	Annunciator	W	Linen/Storage
C	Washstand	M	Secretary's Desk	X	Electric Reg. Locker
D	Ice Box	N	Bookcase	Y	Shower
E	Sink	O	Buffet	Z	Shelves Overhead
F	Drainboard	P	Locker	AA	With Upper Berth
G	Table	R	Refrigerator	BB	Folding Washstand
H	Switchboard	S	Dresser	CC	Folding Hopper
I	Chime	T	Range	DD	Radio
J	Speedometer	U	Water Heater	EE	Convertible Sofa

Business Car No. 18

Drawn as in service at Topeka, August 1967

Floorplan data 23 C-168 rev. C July 24, 1930

Drawn by Frank M. Ellington

Business car No. 18 was one of the Barney & Smith Dining cars, converted to an official car in 1907. During its early days as a wooden Business car, the 18 was assigned to the General Manager, Eastern Lines at Topeka. The 18 stands unassigned at Topeka 53 years later on December 5, 1960 just one year before being retired from official service. The car was now over 70 years old. It was converted to Maintenance of Way Bunk car 194021 at the Topeka shop in December 1961 and was sold for scrap a decade later. *-F.O. Kelley photo, John B. McCall collection*

19
A.T.&S.F.

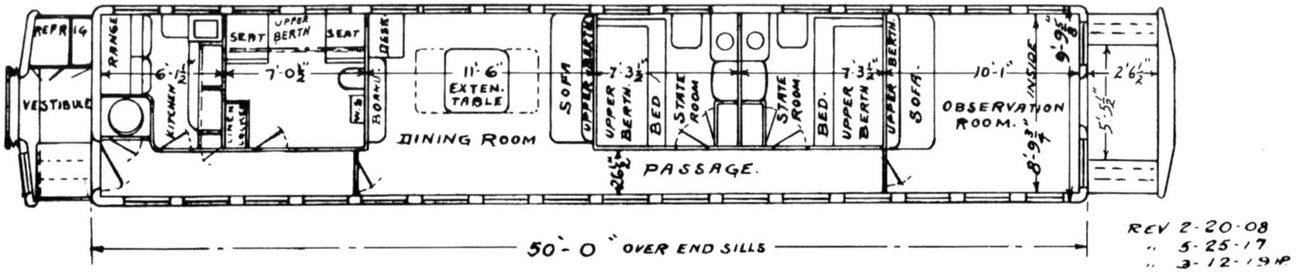

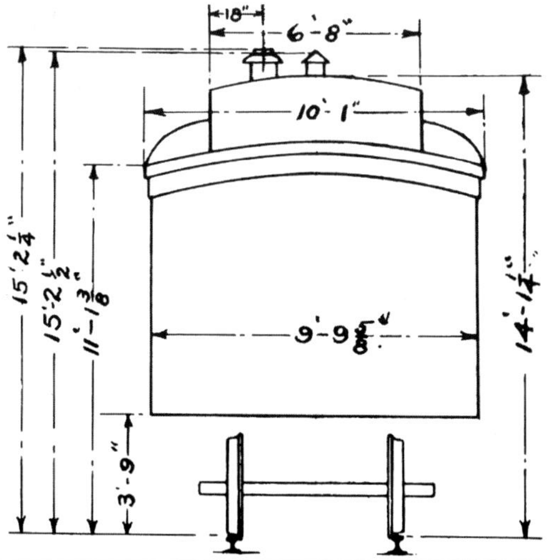

Business Car No. 19

(originally PV&NE 14, PV&NE 1400, PV&NE 15)

AT&SF Business car No. 19 was originally Pecos Valley & Northeastern Business car 14. It was renumbered PV&NE 1400 in July 1899, and renumbered PV&NE 15 in July 1904, before becoming AT&SF 19 in April 1908. This 50-foot car had eight single and two double beds with a sleeping capacity of 12. It was rebuilt with a 6-inch "I" beam underframe and steel platforms and had a light weight of 108,600 lbs. It was dismantled December 30, 1935. No photos have been uncovered of this car.

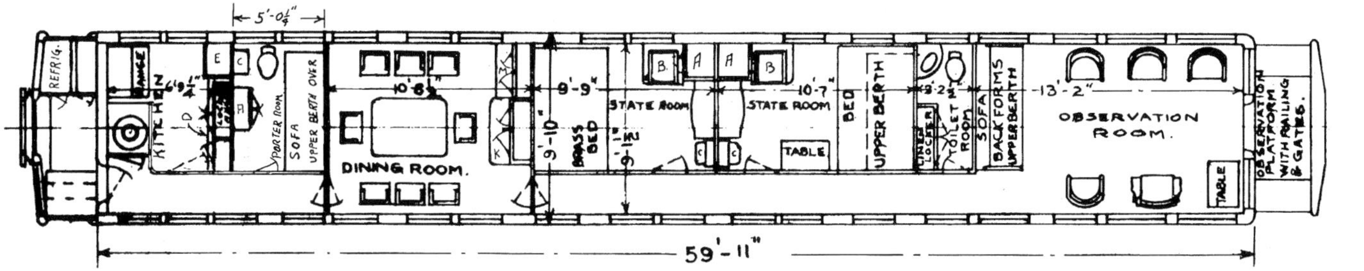

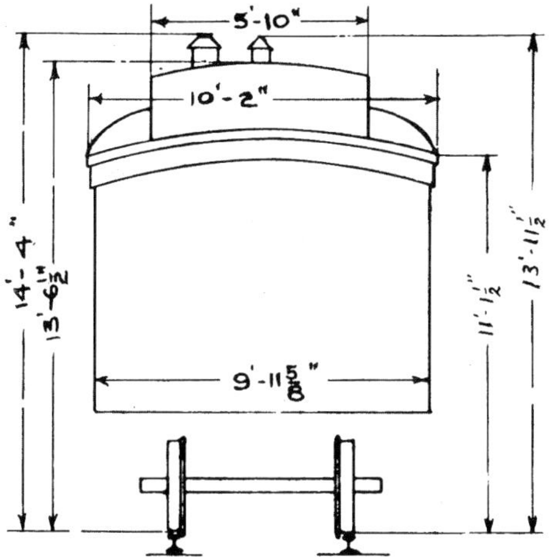

Business Car No. 20

(originally GC&SF 217, GC&SF 12)

B&S 1886

AT&SF Business car No. 20 was originally Gulf, Colorado & Santa Fe Business car 12, built by Barney & Smith in 1886. It was renumbered AT&SF 20 in 1910. This 60-foot car had five single and two double beds with a sleeping capacity of 9. It was rebuilt with a 6-inch "I" beam underframe and steel platforms and had a light weight of 114,200 lbs. It was converted to AT&SF 20, Track Recording Car, in July 1927.

ATSF 20 as Track Recording Car at Topeka, Kansas, March 21, 1937. The 20, having never received steel plate side sheeting or more modern trucks, retained much of its 19th century heritage, including its upper window sash. The car was eventually dismantled March 29, 1940. For more information on car No. 20 refer to Chapter 8. *-W. A. Gibson Sr. Photo, John B. McCall collection*

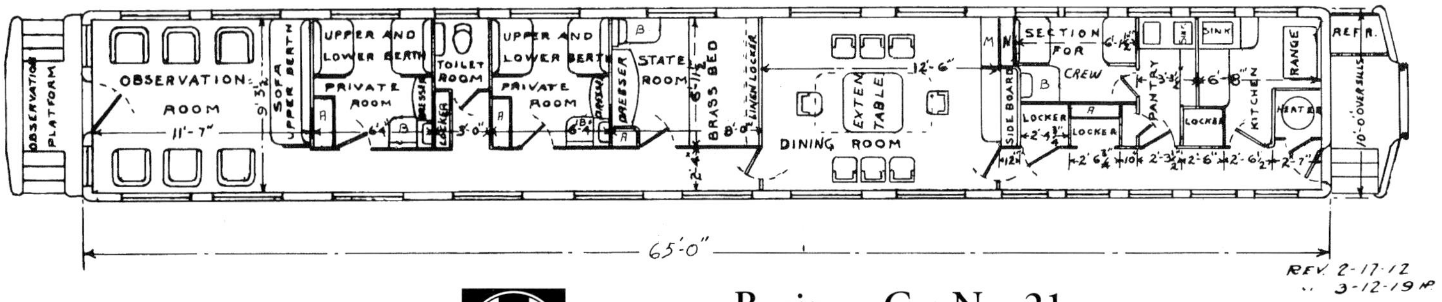

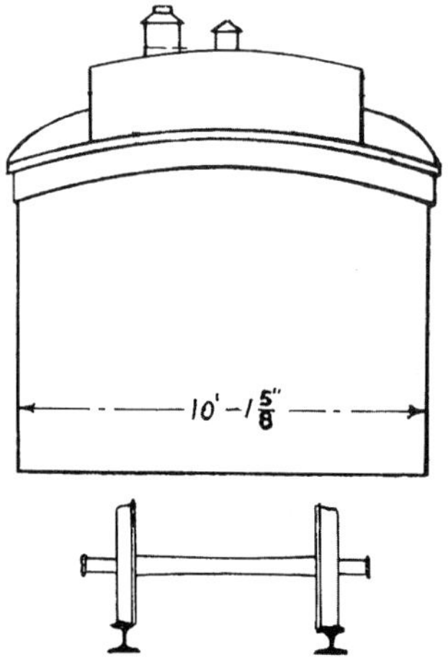

Business Car No. 21
(converted from Diner 1407)
B&S 1890

AT&SF Business car No. 21 was originally ATSF Diner 1407, built by Barney & Smith in 1890, and was converted to Business car AT&SF No. 21 in 1910. This 65-foot car had six single and two double beds with a sleeping capacity of 12. It was rebuilt with a 6-inch "I" beam underframe and steel platforms and had a light weight of 114,100 lbs. It was converted to AT&SF 21, Radio Test Car sometime in the 1940s. Here it is seen at San Bernardino, California, on the Los Angles Division March 18, 1949, in company with another Business car. The apparatus on the roof suggests that No 21 was already being used as a radio test car. No. 21 was eventually retired in 1958 and sold to the Stockton Terminal & Eastern railroad and converted to stationary offices. *-Jack Whitmeyer photo, SFRH&MS Collection*

22
A.T.&S.F.

Business car No. 22 stands ready for assignment to the General Manager, Western Lines at Amarillo, Texas after being upgraded with a steel underframe and bulkheads in addition to steel plating at Topeka February 29, 1928. *-Santa Fe photo, Frank Ellington collection*

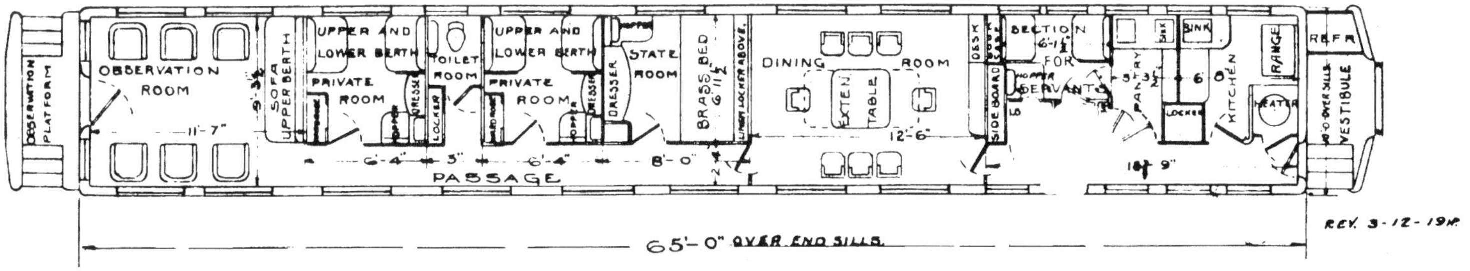

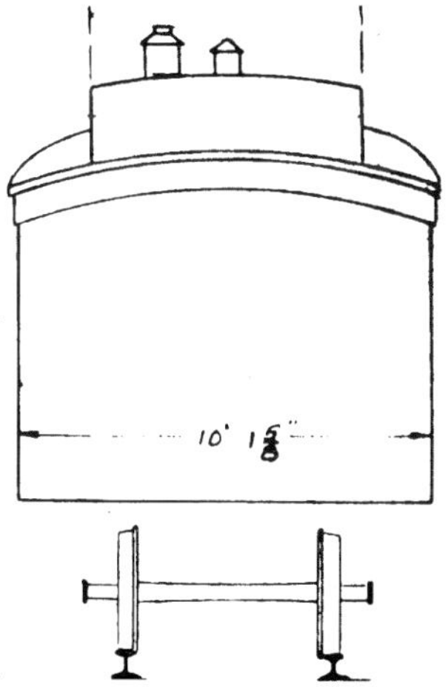

Business Car No. 22
(converted from Diner 1405)
B&S 1890

AT&SF Business car No. 22 was originally ATSF Diner 1405, built by Barney & Smith in 1890. It was converted to Business car AT&SF 22 in 1910. This 65-foot car had six single and three double beds with a sleeping capacity of 12. It was rebuilt with a 6-inch "I" beam underframe and steel platforms and had a light weight of 140,100 lbs.

Rear view of AT&SF Business car No. 22 after being outshopped at Topeka February 29, 1928. *-Santa Fe photo, Frank Ellington collection*

Topeka Shops renumbered Business car No. 22 to 199075 in July 1952 to make room for the heavyweight Business car which had just been converted from Coach 3360. The Business car numbered 199075 was then converted to Maintenance of Way Living car 199075 at Cleburne, Texas, in May 1953, then changed to a Foreman car at Cleburne shops in April 1955. The car was eventually sold for scrap in December 1972. *-Ed Birch collection*

23
A.T.&S.F.

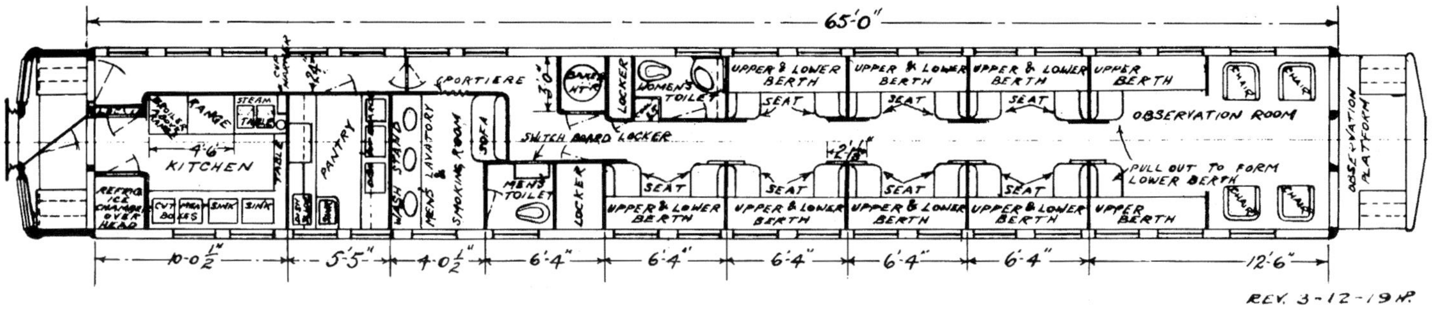

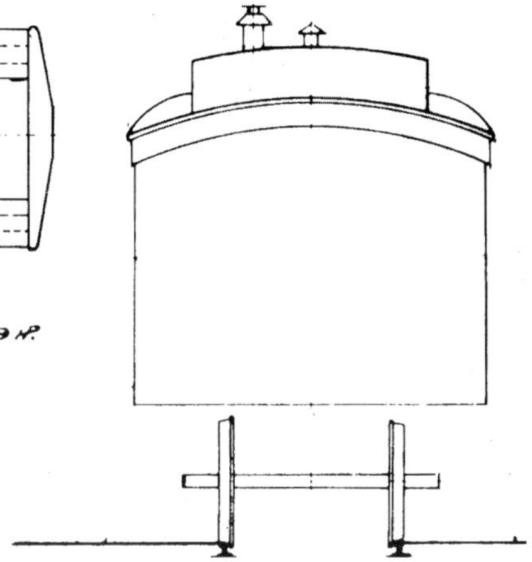

Business Car No. 23
(converted from Diner 1400)
Pullman 1888

AT&SF Business car No. 23 was originally ATSF Diner 1400, built by Pullman in 1888. It was converted to Business car AT&SF 23 in 1913. This 65-foot car was unique among Business cars, as it had nine sections with a sleeping capacity of 18. It was rebuilt with a 6-inch "I" beam underframe and steel platforms and had a light weight of 138,600 lbs. It was assigned as a "Demonstration Car" circa 1934. No photos or other information has surfaced on this car.

24
A.T.&S.F.

AT&SF 24 Business car No. 24 was originally GC&SF Coach 2810 built by Barney & Smith in 1885. The 45-foot, eight wheel car was converted to AT&SF Business car No. 24 in, 1914, but immediately renumbered in May the following year as ATSF 199211, and used as the Corwith, Illinois, Pay car. A 2nd No. 24, used as a Steam Heat Instruction Car, was converted from 70-foot Diner 1412, built by Barney & Smith in 1900, in October 1914. This in turn was converted to ATSF 789, Coach Smoker, in January 1926 (see Special Purpose cars, Chapter 8.) No photos or other information has been discovered on these two cars.

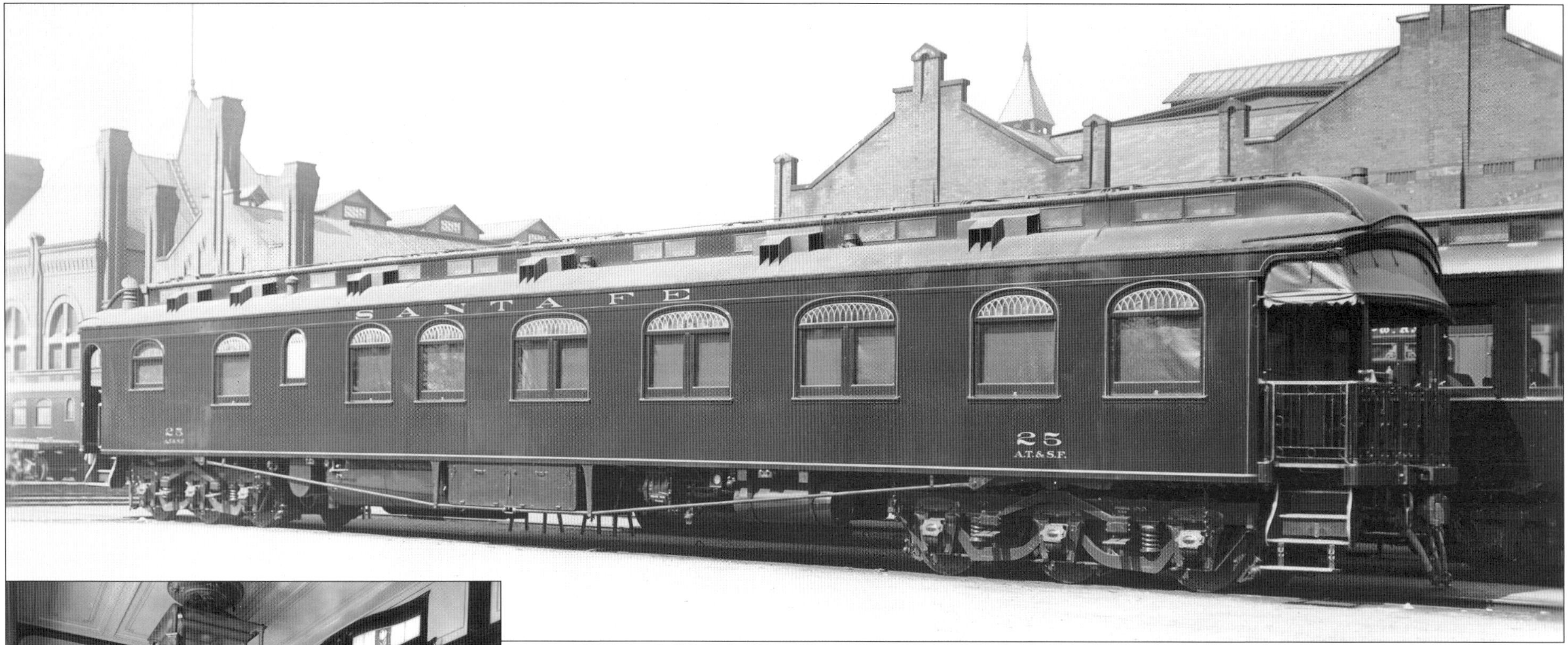

AT&SF car No. 25 was actually a private, or personal car, rather than a Business car in the traditional sense. The 72-foot 12-wheel car was built by the Pullman Company on plan 2299-E dated September 16, 1907, and assigned to Mr. Benjamin P. Cheney, Jr., a Santa Fe director. Its exterior treatment and interior appointments were very much the equal of President Ripley's car No. 17, which was released from Pullman earlier in 1907. Note the stained glass tulips on the inside of the transom windows verses the pattern on the exterior transoms in the observation room of No. 25." Records indicate this car was sold to Texas Delta Farms Co. January 17, 1920, which means it was on the Santa Fe less than thirteen years, very odd for such an elegant car. *-Pullman photos September 30, 1907, courtesy Smithsonian Institution*

SANTA FE 400 SERIES WOOD SUPERINTENDENTS BUSINESS CARS

The Atchison Topeka & Santa Fe was one of the few railroads to possess a fleet of specially-built Business cars for their Division Superintendents. In addition to the previously listed (mostly 12-wheel) wood Business cars in the 1903 series—cars No. 1 through 25—Santa Fe also rostered, in a new 400 series, smaller, and in some cases much older, Business cars used primarily for Division Superintendents. The first cars listed in the 400 series came in 1903-1907, then several more were added in 1923-1925.

Six were renumbered from the wood Business cars in the 200 series in use from the mid-1890s forward. They were the 400, formerly the 213; the 402, formerly the 209; the 403, formerly the 211; the 407, formerly the 210; the 408, formerly the 214 and the 410, formerly the 212. Among these were two of the earliest cars, the 402 and the 407, both of which started life as Chicago Burlington & Quincy Coaches, built in 1870. Six were converted from Way cars. A number were converted from Coaches. Several were early short Business cars owned by Santa Fe or its various subsidiaries. Unique among the 400 series was the No. 419, the former Business car Breckenridge of the Denver, Enid & Gulf Railway, which came with Santa Fe's purchase of that road in 1907.

The 400 series Business cars were generally very short with four-wheel trucks, except 2nd 412, 2nd 413 and the 419, all of which had six-wheel trucks. The shortest car was but 30 feet, the longest 56 feet—this had formerly been AT&SF 10, listed among the large Business cars—but most were in the range of 40 to 45 feet.

Assigned to Division Superintendents around the system, these cars received less attention, perhaps, than their six-wheel counterparts, which carried the President and Directors of the railroad, among other high dignitaries. One car, the 404, had some notoriety, however, serving in the Farm Demonstration Train in the mid-1920s.

For some reason, there was much renumbering of cars in this series, some of this occurring later to make way for new steel heavyweight Superintendent's cars coming on line beginning in 1924. With the new steel cars arriving, a number of these old wood cars still serving in the mid-1920s, were converted to Maintenance of Way Bunk cars. Some were converted to Drovers cars. The 414 was converted to Power Plant Test car 5002, and 2nd 412 to 5013, Rules Examiner car, both in 1927.

All of these cars but two were eventually scrapped. One car, possibly No. 415, wound up as a yard shanty in use at Cleburne through the 1960s. Two of the 400 series wood Superintendent's Business cars were preserved. The 410, which had been built by the Santa Fe originally as Coach 97 at Topeka in 1880, converted to AT&SF 212, Business in December 1893, and renumbered 410 ten years later, had been converted to Bunk car 199171 in May 1929, and was eventually retired in 1949. It was retained by Santa Fe, restored as ATSF 2, in the Cyrus K. Holliday exhibition

Continued on page 73

Superintendent's Business car No. 422 was converted from Coach 2854 at Topeka about 1913. While it was assigned to the Eastern Division, at Topeka, most of its years, it carried three different business car numbers: 422 (1913-1923); 2nd 401 (1923-1925); and 2nd 411 (1925-1927). It was renumbered to make room for the steel Superintendent's cars being delivered. The car, originally built in 1879, was converted to Bunk car 199155 in 1927 and eventually dismantled in 1938. *-Frank Ellington collection*

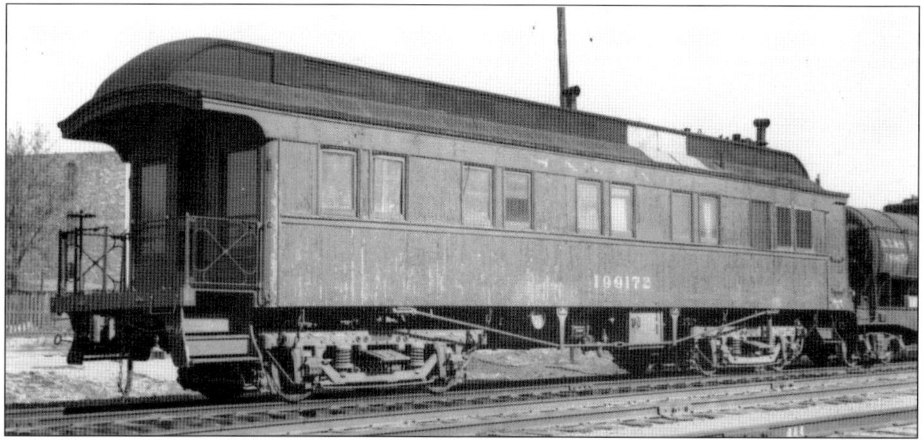

This former Coach, built by the CB&Q in 1870, was purchased second hand by the Santa Fe and converted to a Superintendent's Inspection car in 1901. The car was renumbered several times; from 210 to 407, Superintendent's Inspection car, and assigned to the Rio Grande Division in 1905, and in 1926 it was renumbered to 2nd 420 to make room for the all-steel Division Superintendent's car being delivered. The 420 was converted to a Bunk car in 1928 and renumbered 199172. It is seen March 24, 1938 at Topeka. *-Ray Hilner*

SANTA FE 400 SERIES WOOD SUPERINTENDENT'S BUSINESS CARS

Business Car No.	*Builder*	*Date Built*	*Length*	*Beds**	*Notes*
AT&SF 400	AT&SF, Topeka	1887	46'	3/6	Originally AT&SF 97 Coach[1]; reno. AT&SF 213 1/96; reno. AT&SF 400, 1903; to 199115 Bunk car 9/13/24.
AT&SF 401	Wells & French	1881	30'		Converted from AT&SF 105 Way Car; to MofW 195941, 1919.
AT&SF 2nd 401	J&S	1879	46'	2/2	Renumbered from AT&SF 422, 1923; reno. 2nd AT&SF 411, 11/24/25.
AT&SF 402	CB&Q Ry.	1870	44'	2/5	Probably from Coach to AT&SF 209, Business, in 1901; reno. AT&SF 402, 1903; to 199151, Bunk, 1/25, dismantled 1934.
AT&SF 403	CB&Q Ry.	1871	45'	2/5	AT&SF 300, Coach, reno. AT&SF 211, 1/00; reno. AT&SF 403, 1903; reno. AT&SF 430, 11/25.
AT&SF 404	Quedley Car Co.	1872	36'	2/4	Converted from AT&SF 9, Way Car; assigned to Farm Demonstration Train circa 1924; to 199153 Bunk car 1926.
AT&SF 405	AT&SF, Topeka	1893	34'	2/4	Converted from AT&SF 386, Way Car.; dismantled Topeka 9/25.
AT&SF 406					Converted from AT&SF 185, Way Car; out prior to 1906.
AT&SF 2nd 406	Carter Bros.	1882	41'	2/3	Converted from AT&SF 2582, Coach; converted to 199152 Bunk car 2/8/26; dismantled 6/48.
AT&SF 407	CB&Q Ry.	1870	44'	2/5	Probably converted from Coach to AT&SF 210, Business, 1901; reno. AT&SF 407, 1905; reno. 2nd AT&SF 420, 1926.
AT&SF 408	-	-	-	-	AT&SF 63, Coach to AT&SF 63, Official, to AT&SF 98, 1888-90; reno. AT&SF 214, 10/93; reno. AT&SF 408, 1903[2]; out prior to 1907.
AT&SF 2nd 408	Barney & Smith	1871	44'	2/5	Converted from AT&SF 2504, Coach; to 199170, Bunk, 1928; dismantled 1938.
AT&SF 409	AT&SF, Topeka	1892	34'	2/4	Converted from AT&SF 325 Way Car; sold to the Missouri & North Arkansas Ry. 5/1/25, reno. 100 Business.
AT&SF 410	AT&SF, Topeka	1880	41'	1/5	AT&SF 97, Coach[1]; to AT&SF 212, Business, 12/93; reno. AT&SF 410, 1903; to 199171, Bunk 5/29, retired 5/49, to Exhibit No. 2.
AT&SF 411	AT&SF, Topeka	1881	42'	2/5	Santa Fe Pacific 96 Officers' car reno. AT&SF 411, 7/02; to 199049 Bunk car 11/24.
AT&SF 2nd 411	J&S	1879	46'	2/2	Renumbered from AT&SF 2nd 401; converted to 199155, Bunk car, 1927; dismantled 1938.
AT&SF 412	AT&SF, Topeka	1897	42'	2/5	Santa Fe Pacific 95 Officers' car, reno. AT&SF 412, 7/02; converted to 199125 Bunk car 6/25, dismantled 1934.
AT&SF 2nd 412	H&H	1882	51'	2/7	Reno. from AT&SF 9, Business; converted to AT&SF 5013 Rules Examiner car 4/27.
SF&SJV 413	Carter Bros.	1902	41'	2/4	SF & SJV Ry. 94, Officers' car; reno. SF&SJV 413, 1905; reno. AT&SF 413,1906; converted to 199050 Bunk car.
AT&SF 2nd 413	A&P RR	1883	56'	2/5	Reno. from AT&SF 10 Business.
AT&SF 414	So. Cal. Ry.	1898	41'	2/2	Southern California Ry. 97, Officers' car, reno. AT&SF 414, 6/04; converted to AT&SF 5002, Power Plant Test car, 1927.
GC&SF 415		1886	45'	2/4	Converted from GC&SF 15, Coach, 1902; to GC&SF 105, Supt's Business; reno, GC&SF 415, 1904; to AT&SF 910, Drovers car, 1927; dismantled 9/34.
GC&SF 416			40'		Converted from GC&SF 6, Coach, 1901; to GC&SF 104, Supt's Business; reno. GC&SF 416, 1900.
GC&SF 2nd 416		1881	45'	2/5	Converted from GC&SF 2562, Coach; converted to GC&SF 9138, Bunk car.
GC&SF 417	AT&SF, Topeka	1890	30'		Converted from GC&SF 58, Way Car.
AT&SF 2nd 417	AT&SF, Topeka	1880	44'	2/5	Converted from AT&SF 2500, Coach; converted to 199161, Bunk car, 1928.
GC&SF 418			45'	2/6	Converted from GC&SF 5, Bagg/Mail/Exp to GC&SF 103, Official; reno. GC&SF 418, 1905; dismantled 10/25.
AT&SF 419		1898	50'	2/5	DE&G Officers' car BRECKENRIDGE[3]; reno. AT&SF 419, Supt's Business, 1907; converted to 199174, Bunk, 3/30; dismantled 1947.
GC&SF 420			40'	2/4	Converted from AT&SF 2583, Coach; converted to GC&SF 9093, Bunk car, 11/24.
AT&SF 2nd 420	CB&Q Ry.	1870	44'	2/5	Renumbered from AT&SF 407, 1926; converted to 199172 Bunk car 1928.
AT&SF 421			45'	2/5	Converted from GC&SF 2564, Coach; to AT&SF 911, Drovers, 1927; to 199175, Bunk 3/35; retired 4/56, to Exhibit No. 3.
AT&SF 422	J&S	1879	46'	2/5	Converted from AT&SF 2845, Coach; reno. 2nd AT&SF 401, 1923.
AT&SF 430	CB&Q Ry.	1871	45'	2/5	Renumbered from AT&SF 403 11/25; out between 1925-1927.

*All four wheel except 2nd 412, 2nd 413 and 419. * Beds: double/single*

Note 1. Early 1890s records list two coaches numbered AT&SF 97, both converted to Official cars numbered 97. One was renumbered AT&SF 212, 12/93, the other was renumbered AT&SF 213 in 1/96.

Note 2. Superintendent's Business car 408 was leased to the SF&SJV RR.

Note 3. The Denver, Enid & Gulf was purchased in 1907 and with it came D.E.& G. Business car BRECKENRIDGE, which became AT&SF Superintendent's Business car 419.

Continued from page 71
train. The 421, a nineteenth century car of undetermined origin, had once been GC&SF Coach 2564 before entering Superintendent's car service. It was converted to Drover's car 911 in 1927, to a Maintenance of Way Bunk car 199175 in March 1935, and eventually retired in 1956. It too was retained, restored, and renumbered AT&SF 4 for the Cyrus K. Holliday exhibition train.

SUPERINTENDENT'S BUSINESS CAR ASSIGNMENTS

Car No.	*Division Assignment* *1907*	*1919*	*1927*
400	Missouri	Missouri	-
401	Eastern	-	-
402	So. Kansas	So. Kansas	-
403	Middle	Middle	-
404	Oklahoma	Oklahoma	-
405	Western	River	-
406	Colorado	-	-
2nd 406	Colorado	Slaton	-
407	New Mexico	Rio Grande	-
408	Rio Grande	-	-
2nd 408	Rio Grande	Western	Unassigned
409	Panhandle	Colorado	-
410	Illinois	Illinois	Unassigned
411	Albuquerque	Albuquerque	-
412	Arizona	Arizona	-
2nd 412	-	-	Arizona
413	Valley	Valley	
2nd 413	-	-	Colorado
414	Los Angeles	Los Angeles	Unassigned
415	Northern	Northern	-
416	Southern	-	-
2nd 416	Southern	Southern	-
417	Unassigned	-	-
2nd 417	-	Plains	Unassigned
418	Beaumont	Beaumont	-
419	-	Pecos	Panhandle
420	-	Galveston	-
2nd 420	-	-	Unassigned
421	-	Panhandle	Western
422	-	Eastern	-
430	-	-	Middle

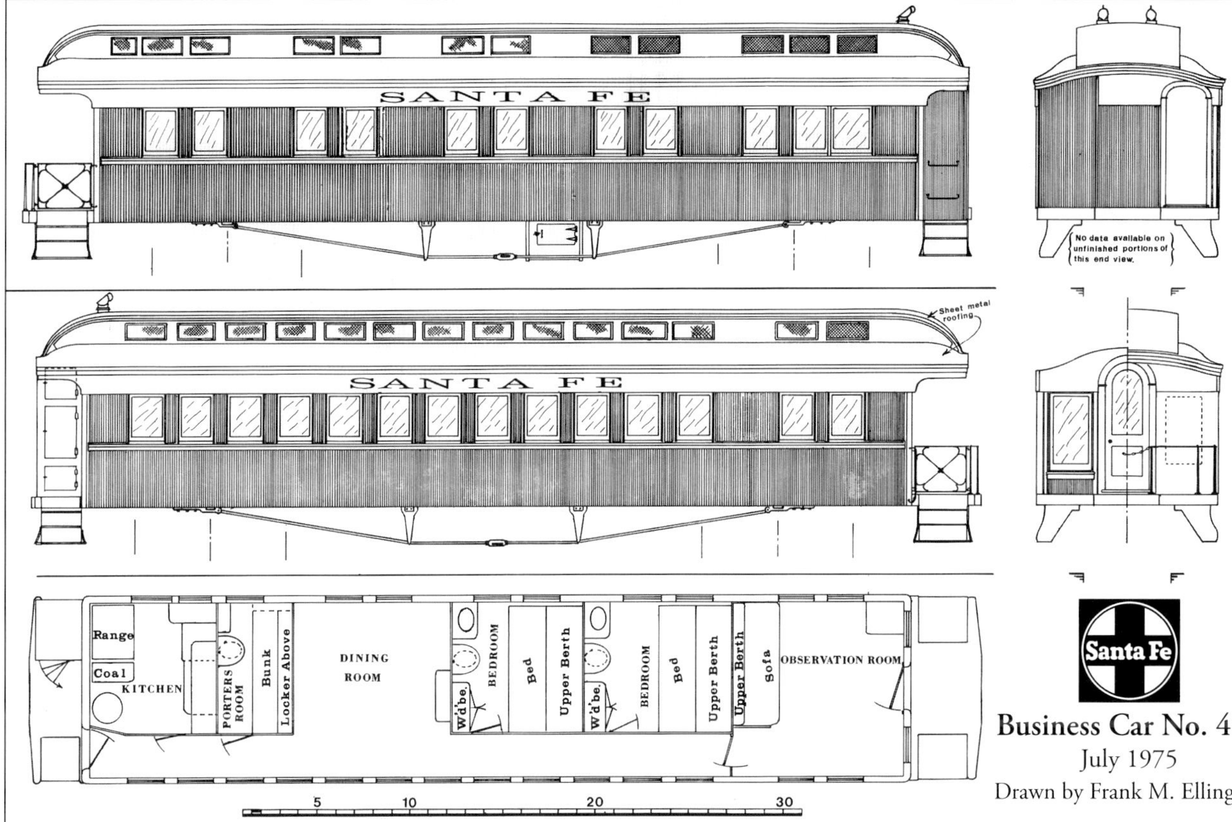

Business Car No. 407
July 1975
Drawn by Frank M. Ellington

Originally built by Wells & French as Way car AT&SF 105 in 1881, this 30-foot car was converted to Superintendent's Business car 401, where it seen at Quenemo, Kansas, in the early-1900s, while assigned to C. T. McLellan, Superintendent of the Eastern Division. The car was converted to MofW 195941 in 1919. *-Santa Fe Collection, Kansas State Historical Society*

In company with an old Coach, Superintendent's Business car 2nd 406 is spotted at Dodge City, Kansas, above, sometime after 1906. Built by California car builder Carter Bros., this car came to the Santa Fe from the Southern California Railway as Coach 102 and was renumbered AT&SF 2582. It was converted to a Superintendent's Business car in 1906 and numbered 2nd 406. It was assigned to the Colorado Division and later the Slaton Division before being relegated to Maintenance of Way Bunk service in February 1928 as 199152. It is seen, left, at Stillwater, Oklahoma, on December 11, 1937, and would continue in this service until June 1948. -*Two photos, John B. McCall collection*

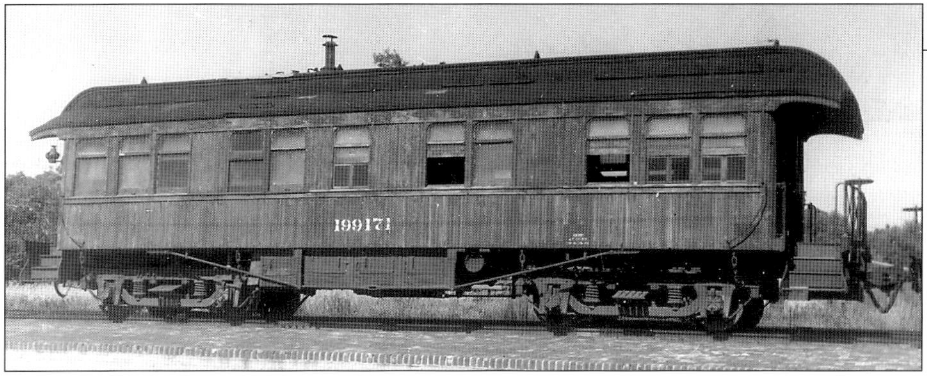

This car, originally constructed by Topeka Shops in 1880, served as Santa Fe Coach 97 until the early 1890s, when it was converted to Superintendent's Business car No. 97. It was renumbered 212 in December 1893, and AT&SF 410 in 1903. As 410 it served as the Superintendent's car for the Illinois Division until May 1929, when it was converted to Bunk car 199171. In the view at left during the busy war years, it was being used as a caboose on extra 3105 West at Stillwater, Oklahoma, on August 8, 1942. It was retired in May 1949. It was retained by the Santa Fe, however, restored and converted back to Coach configuration, numbered "2" and became part of the Cyrus K. Holliday exhibit, in which service it is seen, right, at Dallas in October 1965. *-Left, John B. McCall collection, right, John B. McCall photo*

What is believed to be the former car 415, built in 1888, was serving as a yard shanty at Clebune, Texas, in 1969. *-Two photos, Frank Ellington*

Exhibit "3" was originally constructed during the 1880s as GC&SF Coach No. 14. It was renumbered GC&SF 2564 Coach in 1902, and was converted to Superintendent's Business car 421 in 1911. In 1927 it was converted to Drover's car 911, and in 1935 to Bunk car 199175. It was subsequently converted back to a Coach for the exhibit train. Dallas, October 1965. *-John B. McCall*

Built in 1872 by the Quedley Car Co., as a Way car, as a Business car assigned to the Oklahoma Division numbered 404, it slept eight within its modest confines. In 1926 the 404 was converted to Bunk 199153 and is seen here at Topeka in 1948. *-Ray Hilner*

Santa Fe heavyweight Business cars 38, built in Pullman lot No. 4883 in 1925; (2nd) 20 built by Pullman in 1928 as Pullman Private car Henry Stanley, which Santa Fe acquired in 1944; and car 39, built in Pullman lot No. 6165 in 1928, are parked in front of the distinctive Santa Fe depot at San Diego, California, in the 1940s. All of these cars served the Santa Fe for another four decades in one form or another, except car (2nd) 20, which was sold in January 1971. Of the three cars, (2nd) 20 has yet to receive air conditioning as the massive duct work is absent from the roof.
-R.P. Middlebrook photo, Stan Kistler collection

4

Heavyweight Business Cars

At the close of World War I, it had become evident that Santa Fe's fleet of rebuilt wood Business cars was showing its age. The greater percentage of these cars had originally been built prior to 1890, and the oldest, No. 9, had been built in 1882 and was nearly 36 years old. Additionally, the war had taken a heavy toll on equipment and Santa Fe was interested in acquiring some modern heavyweight cars.

Santa Fe's heavyweight Business cars came in three distinct groups. The first, with one exception, were ordered by and delivered to the Santa Fe in Business car configuration direct from the factory. These cars were all constructed by the Pullman Company within a ten-year span beginning in 1918 as follows: No. 30 in Pullman lot No. 4530 in 1918, and USRA 82 (to become No. 31 in 1920); Nos. 32-34 in Pullman lot No. 4727 in 1923; Nos. 35 and 36 in Pullman lot No. 4794 in 1924; Nos. 37 and 38 in Pullman lot No. 4883 in 1925; Nos. 2nd 9 and 2nd 10 in Pullman lot No. 6113 in 1928; and No. 39 in Pullman lot No. 6165 in 1928. The accommodations contained in the 30-series cars amounted to five double beds and seven single beds, while cars Nos. 9 and 10 could accommodate four double beds and seven single beds.

As World War II was drawing to a close, Santa Fe acquired two additional heavyweight Business cars. The first was 1927-built Pullman Private car Henry Stanley, which Santa Fe acquired in 1944 and renumbered to No. 2nd 20. The second was 1927-built Pullman Private car Edgewood, which Santa Fe acquired in January 1945 and renumbered 40, but also named Santa Fe.

The final group of three heavyweight Business cars were all converted from 70-foot 6-inch Coaches at Topeka in 1952. Originally to be numbered 41, 42 and 43, the numbers were changed to 4th 3, 2nd 17 and 2nd 22 at the time they were outshopped to fill vacancies in the wooden car roster.

All of these heavyweight cars received Waukesha air conditioning between 1947 and 1953 with the distinctive roof duct modifications. Apparently all the 30 series cars were fitted with roller bearings in the late 1950s and early 1960s, and most received new trucks using straight equalized trucks from late-delivery heavyweight head end cars, which were then fitted with outside swing bolsters, Decelostats and truck-mounted brake cylinders; some may have received disk brakes. Car No. 39 (and maybe the 38) was delivered new with the straight equalized trucks.

All of these heavyweight Business cars were originally painted the standard green. During the late-1950s and early-1960s, all, with the exception of the 2nd 9, 2nd 10 and the 30, were painted silver and shadow-lined. Car 31 through 36 and the 39 were rebuilt with Thermopane windows. Car 38, after being renumbered 58 in 1973, received Thermopane windows as well. Cars 31, 32, and 35 through 39 had their roofs rounded and fully enclosed for the full length of the car leaving no trace of ever having had a monitor type roof.

Car No. 34 was the first in this class to be air conditioned in 1946, then car 39 in 1947 and the remaining cars in 1950. This whole series received Waukesha-type air conditioning.

During the most of their service lives, each of the three General Managers were assigned a heavyweight Business car, as well as five of the six Assistant General Managers. Cars 2nd 9 and 2nd 10 were the exception. These were assigned as Superintendent's cars; 2nd 9 was assigned to Winslow for

Built by Pullman in 1928, heavyweight Business car 2nd 9 is steamed up and ready for occupancy at Topeka about 1954. *-Ed Birch collection*

Most of Santa Fe's heavyweight Business cars were delivered with six-wheel bolted-pedestal Pullman trucks.

Cars 38 and 39 were delivered with straight equalized trucks, and as the other cars were modernized, most were fitted with these trucks from late-delivery heavyweight head end cars that were being retired.

years. By the early 1970s, there was only one car assigned to each General Manager's territory, with one exception. When the GC&SF was merged into the AT&SF, the General Manager at Galveston went to Chicago as the Assistant Vice President, Personnel. His assistant went to Amarillo as the second Assistant General Manager for the Western Lines. This A.G.M. had jurisdiction over what was formerly the GC&SF (Northern and Southern Divisions), but being so far off his territory in Amarillo, one Business car was kept in Fort Worth, so on the Western Lines there were two Business cars for the G. M. and his two assistants.

Several of these heavyweight cars were sold or retired from Business car service during the 1960s and 1970s. Car 30 was converted to a Film car in March 1969, and car 39 was chosen to become a Track Geometry car in 1973.

These cars were normally operated in passenger trains. In the event a car was to be used behind one of Santa Fe's fast freights, these Business cars were operated in conjunction with one of the Steam Generator cars, as they required steam to actuate the heating and air conditioning systems. The cars which remained in Business car service in 1973, Nos. 31, 32, 34, 35 and 38, were renumbered 54 through 58. At this time, these remaining cars were shopped and equipped with self-contained power plants making the use of a Steam Generator car unnecessary.

The 32 and 34 were sold to Conrail, with the 34 being used for parts, the 35 was sold as is now in private car service. By the late 1980s, this left Office car No. 58, the former 38, built in 1925 as the only Pullman heavyweight Business car in service on the Santa Fe.

During the Mike Haverty era, one of the former General Manager's Pullman heavyweight Business cars, No. 54, the former 31, was moved to Topeka Shops for refurbishing on November 6, 1989. The car reappeared Februrary 29, 1990, remodeled and renumbered 1869 and named CYRUS K. HOLLIDAY. Interestingly, the Vice President of Operations at that time, had come from the UP, was the one responsible for numbering the car 1869. He argued for the number 1869 because it represented the first year of train operations on the AT&SF. It was in May 1869 that the bridge across the Kansas River between North Topeka and Topeka was completed with regular train service south of Topeka beginning later that year.

The 1869 was designated Santa Fe's number one "show car," and today this car is on static display at the BNSF corporate headquarters in Fort Worth, Texas.

Car 35 at Fort Worth in the early 1970s, just prior to renumbering to car 57, exhibits the typical exterior modifications common to the "30 series" cars. *-John B. McCall*

SANTA FE HEAVYWEIGHT BUSINESS CARS

Business Car Nos.	*Builder Date Built*	*Length & Trucks*	*Notes & Disposition*
AT&SF 30	Pullman 1918	73'6" 6 wh	Converted to Film car 30 3/69, reno. 5009 7/70, reno. 73 1/73; sold to Kasten Railcar, 2/8/88; derelict, Cherryvale, Kansas 2007.
AT&SF 31	Pullman 1918	73'6" 6 wh	Built as USRA 82, acquired by Santa Fe 1920 reno. AT&SF 31 Business, reno. AT&SF 54 Business car 2/12/73, modernized and reno. AT&SF 1869 1989; on display at BNSF headquarters, Fort Worth Tx.
AT&SF 32	Pullman 1923	73'6" 6 wh	Renumbered AT&SF 55 Business car 8/19/73 Chicago, sold to Conrail, converted to Track Geometry car CR 31, October 1983.
AT&SF 33	Pullman 1923	73'6" 6 wh	Sold to Atlantic Richfield 11/69 Chicago; passed through several hands; renamed Tamalpais, in private car service, 2007.
AT&SF 34	Pullman 1923	73'6" 6 wh	Renumbered AT&SF 56 Business car 2/12/73 Ft. Worth, sold to G. Merle-Smith, 1982; sold to Conrail 1984 for parts; sold to John Baldwin, moved to Oakdale, Ca., 1984; purchased by Mr. Tom Hoback, president, Indiana Rail Road, 2004, fully restored, 2005.
AT&SF 35	Pullman 1924	73'6" 6 wh	Renumbered AT&SF 57 Business car 1/15/73 Topeka; sold to G. Merle-Smith, 1982; sold to F.L. Mosely, 1983, renamed Louisa; private car service, named Casper Mountain, 2007.
AT&SF 36	Pullman 1924	73'6" 6 wh	Retired 9/72; sold to G. Merle-Smith, 1982; as of this writing reportedly a Denver real estate office.
AT&SF 37	Pullman 1925	73'6" 6 wh	Sold to a private party, 10/70 Ft. Worth.
AT&SF 38	Pullman 1925	73'6" 6 wh	Renumbered AT&SF 58 Business car 2/16/73 LA; was still in service as late 6/11/90; sold to Arizona & California as No. 58.
AT&SF 39	Pullman 1928	73'6" 6 wh	Converted to Track Geometry car 5020, reno. 85 prior to being out-shopped 9/4/73.
AT&SF 40	Pullman 1927	74'9" 6 wh	Private car Edgewood, acquired by Santa Fe named 1st Santa Fe Business car 1/45, again reno. AT&SF 40 Business car 12/49, retired 3/63, sold to El Paso Iron & Metal, Vinton, Tex. 8/24/66.
AT&SF 3rd 4	Pullman 1928	70' 6wh	Rebuilt from AT&SF 3346 Coach 1952[1], reno. AT&SF 21 Business car 7/67.
AT&SF 2nd 9	Pullman 1928	66'9" 6 wh	Retired 9/67, sold for scrap 5/16/68, resold to a private party. Currently part of an Antique mall in Fairfield, California..
AT&SF 2nd 10	Pullman 1928	66'9" 6 wh	Retired 11/62, reinstated and converted to 194450 Bunk car 3/67, reno. 203739 Bunk car 11/28/73, at Saugus, California, 2007.
AT&SF 2nd 17	Pullman 1928	70' 6wh	Rebuilt from AT&SF 3361 Coach 1952[1], sold to Grants Supply Co., Tulsa, Ok. 12/70.
AT&SF 2nd 20	Pullman 1927	83'6" 6 wh	Henry Stanley acquired by Santa Fe 1944 reno. AT&SF 20 Business, sold to Grants Supply Co., Tulsa, Ok. 1/71.
AT&SF 2nd 21	Pullman 1928	70' 6wh	Reno. from AT&SF 4 Business 7/67, sold to Grants Supply Co., Tulsa, Ok. 12/70, wound up in Mexico as car 21 Mazatlan.
AT&SF 2nd 22	Pullman 1928	70' 6wh	Rebuilt from AT&SF 3360 Coach 1952[1], Coach, sold to Grants Supply Co., Tulsa, Ok. 12/70.

1. These three cars were originally to be numbered 41,42 and 43, but these numbers were changed to 4,17 and 22 prior to being out-shopped in 1952.

HEAVYWEIGHT BUSINESS CAR ASSIGNMENTS

Business Car Nos.	*Assignment February 18, 1928*	*February 19, 1954*	*February 8, 1960*	*January 1, 1966*
AT&SF 30	Directors' car	T. A. Blair	R.H. Beeder Chief Engineer System, Chicago	Unassigned
AT&SF 31	General Manager	J. P. Morris	T.T. Blickle General Manager Mechanical Dept., Chicago	Mechanical Dept.
AT&SF 32	Vice President Operating Dept.	W. L. More	W.L. More Vice President, Chicago	Chief Engineer-System
AT&SF 33	Unassigned	G. E. Duffy	J.H. Blake Asst. General Manager, Los Angeles	Asst. General Manager Coast Lines
AT&SF 34	Vice President	Unassigned	J.B. Noe Asst. General Manager, Topeka	Asst. General Manager Eastern Lines
AT&SF 35	Asst. General Manager Eastern Lines	G. R. Buchanan	R.G. Buchanan General Manager, Amarillo	General Manager Gulf Lines
AT&SF 36	V.P. & General Manager GC&SF	J. P. Cowley	O.H. Osborn V.P. & General Manager, Galveston	General Manager Eastern Lines
AT&SF 37	General Manager Eastern Lines	J. N. Landreth	L.M. Olson General Manager, Topeka	Asst. General Manager Gulf Lines
AT&SF 38	General Manager Coast Lines	O. L. Gray	J.N. Landreth General Manager, Los Angeles	General Manager Coast Lines
AT&SF 39	-	C. R. Tucker	C.R. Tucker Vice President Operations, Chicago	Chicago pool
AT&SF 40	-	R. G. Ryden	R.G. Ryden Vice President, Chicago	-
AT&SF 3rd 4	-	E. P. Dudley	J.A. Parkinson Gen.Supt. Communications & Signals	Gen.Supt. Communications & Signals
AT&SF 2nd 9	Unassigned	P. O'Sullivan	O.R. Hammit Superintendent Albuquerque Division, Winslow	Supt. Albuquerque Div.
AT&SF 2nd 10	Unassigned	R. D. Shelton	Unassigned	Unassigned
AT&SF 2nd 17	-	J. E. Lester	T.J. Anderson Asst. General Manager, Amarillo	Asst. General Manager Gulf Lines
AT&SF 2nd 20	-	Unassigned	Unassigned	Chicago pool
AT&SF 2nd 21	-	-	-	-
AT&SF 2nd 22	-	O. H. Osborn	J.E. Lester Asst. General Manager , Topeka	Asst. General Manager Eastern Lines

Car No. 30 brings up the rear of train No. 75 west of Lubbock, Texas, on July 19, 1960, while assigned to the Chief Engineer, System, out of Chicago. *-John B. McCall*

30
A.T.&S.F.

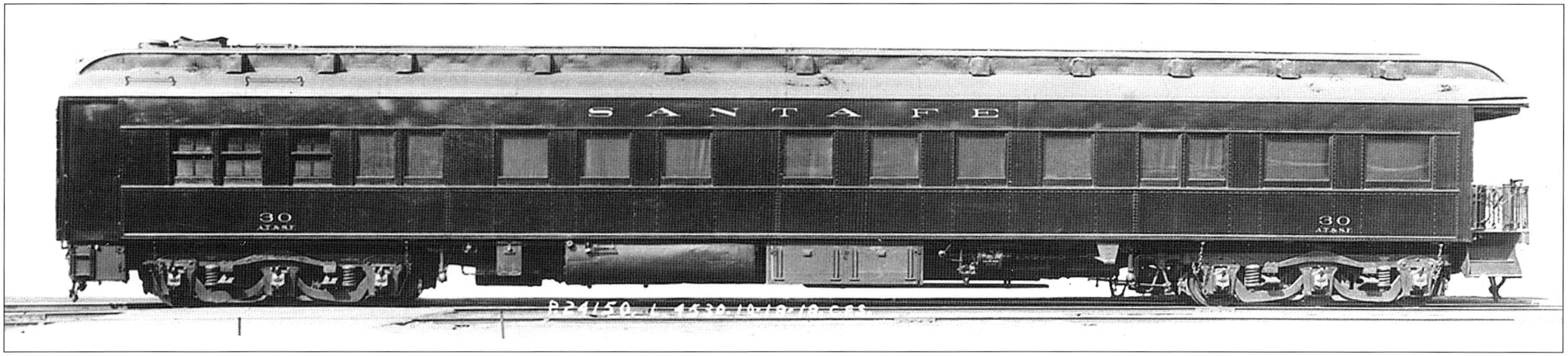

Santa Fe Business car 30, seen in its builder's photo dated October 18, 1918, was the first of ten almost identical cars constructed for use by Santa Fe officials between 1918 and 1928. The 30 was delivered by Pullman in lot No. 4530. There were two cars in lot No. 4530, the other being United States Railroad Administration car 82, which became A.T.&S.F. 31 in 1920. Note the single gold stripe below the initials; this was deleted soon after these cars were delivered. *-Pullman photo, Frank Ellington collection*

Car 30 is attached to No. 75, the *California Special*, at Lubbock in 1959. At that time the car was assigned to R.H. Beeder, Chief Engineer, System, Chicago. The 30 has been modernized and has received Waukesha air conditioning with the distinctive roof duct modifications and SKF roller bearings on the axles in place of the older solid bearings. The car is in green livery and during its life time as a Business car was never painted silver and shadow-lined. *-Ray Hilner photo, John B. McCall collection*

The opposite side of car 30. Here it is spotted in Lubbock July 19, 1960, prior to being added to the consist of train No. 75. *-John B. McCall*

In March 1969, car 30 was converted to a Film car at Topeka. Parts of the platform canopy were cut away, and a special housing built in which cameras and other apparatus for filming sequences for the locomotive simulator were installed. For more information on the Film car, see chapter 8. *-John B. McCall collection*

Business car 30 is seen in service at Fort Worth, Texas, on August 4, 1963. Aside from air conditioning and roller bearings, car 30 was never upgraded as much as the other 30 series cars. Assigned to R.H. Beeder, Chief Engineer, System, out of Chicago in 1960, it was "unassigned" as early as January 1, 1966, and was stored at Topeka in 1968. *-Keith Jordan collection*

After a number of years as a Film car, the 30 (last numbered 73) was retired and sold to Kasten Railcar on February 8, 1988. It was stored for 15 years in Santa Ana, California, and then was moved to Cherryvale, Kansas, where it sits, rather worse for wear in this view from July 2007. *-John R. Signor*

31
A.T.&S.F.

The second Business car in Pullman lot No. 4530 was built in 1918 for the United Stated Railroad Administration and numbered 82. During World War I, car 82 was assigned to one of the Federal Managers, who was actually the president of the Santa Fe. After the war, in March 1920, it was sold to the Santa Fe and renumbered ATSF 31. Pullman Co. photo, October 25, 1918. *-Frank Ellington collection*

Here we see car 31 at Topeka in the early 1950s, with a blue flag applied, after it had its roof rounded and fully enclosed for the full length of the car. Note the radio antenna running on the roof. In subsequent modernizing it will receive an enclosed platform, straight equalized trucks and Thermopane windows. *-F.O. Kelley photo, John B. McCall collection*

Business car 31 has received additional modifications, in these views. The observation platform area is now sheet steel. The car has retained its Pullman trucks and coach green paint scheme, with only the number 31 and the customary "Santa Fe" in yellow. *-Four photos, Santa Fe, John B. McCall collection*

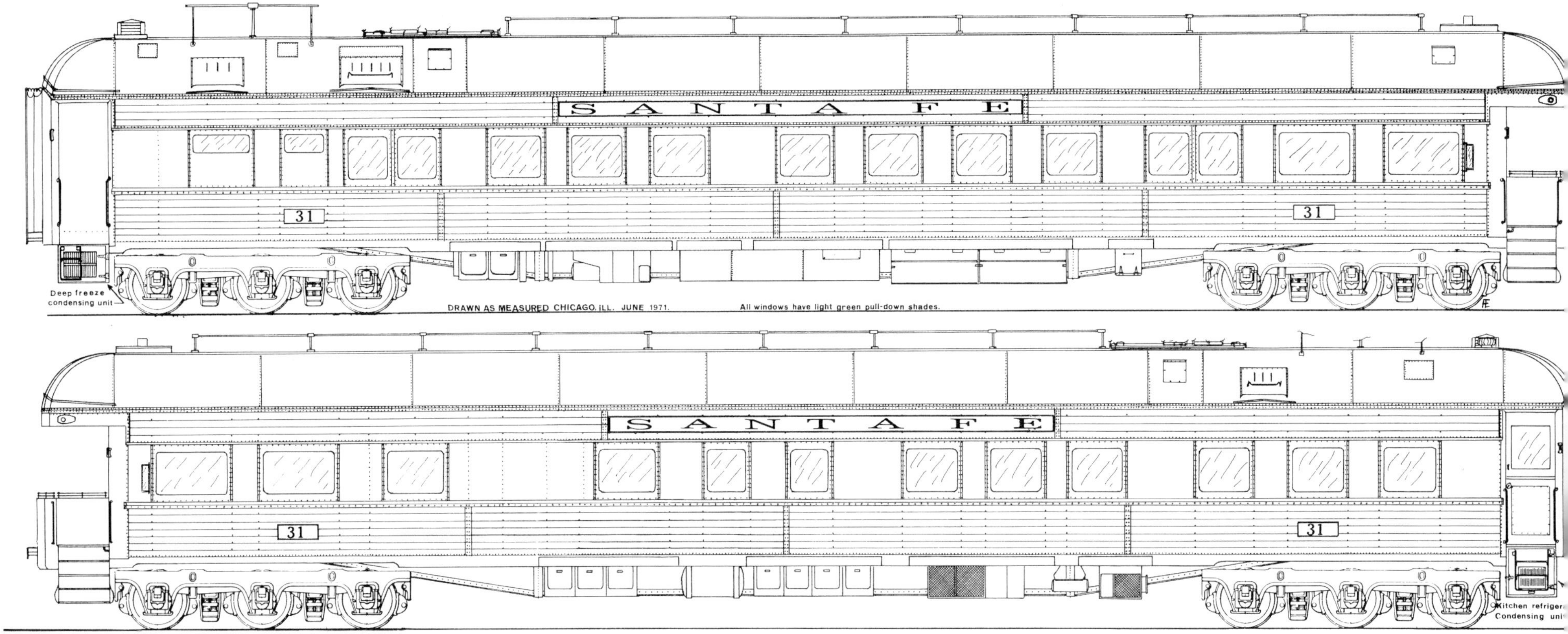

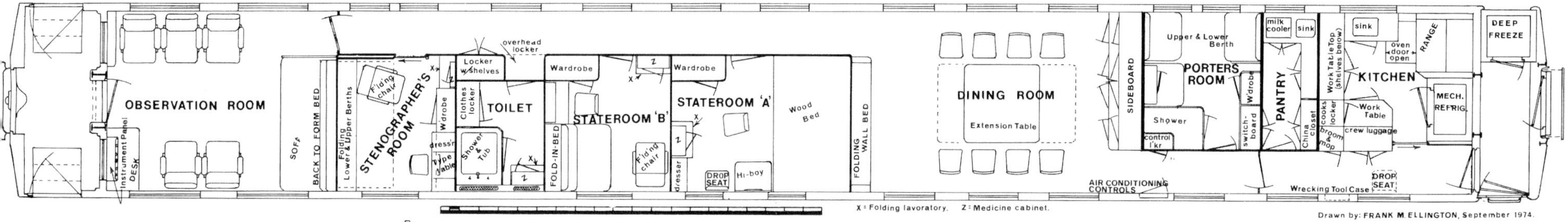

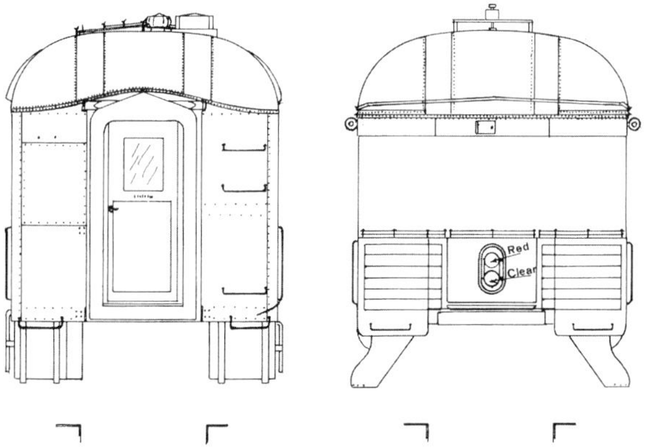

Business Car 31

Drawn as measured
Chicago, Illinois, June 1971
Drawn by Frank M. Ellington

Car 31 was renumbered 54 on February 12, 1973. Now thoroughly modernized with Thermopane windows, fully-enclosed roof and straight equalized trucks, the 54 is seen at Topeka on December 28, 1974 in fading shadow-lined livery. *-Frank Ellington*

By the late 1980s, the last heavyweight Business car in service on the Santa Fe was No. 58, the former 38, built in 1925. It has been modernized on the interior in the 1950s with blonde paneling replacing the rich interior woods originally used by Pullman. During 1989, the 54 was remodeled again with rich polished wood being restored. It was renumbered 1869, named Cyrus K. Holliday, and received a silver paint scheme with a deep Santa Fe blue belt rail. *-All, Frank Ellington collection*

Nearly all of the exterior modifications of the Cyrus K. Holliday in this front ¾ view were completed on the car as No. 54, such as new trucks and Thermopane windows—including the two small galley area windows instead of the original three. *-Frank Ellington collection*

Car ATSF 1869, Cyrus K. Holliday, in service at Barstow in February 1992. This is the rear of the car, but note the stainless steel inserts and temporary diaphragm fitted to the observation platform. Similar inserts were fabricated at Topeka for some of the lightweight cars as well, to make open platform cars suitable for mid-train use under all weather conditions. *-John B. McCall*

In service view of car 1869, Cyrus K. Holliday, at Stockton, California, August 1, 1990. The car had a self-contained power supply, which allowed it to be attached to freight trains without the addition of a Steam Generator car. *-Bill Childers*

As the only heavyweight car in service at the time of the Santa Fe-BN merger in 1996, BNSF chose to place the 1869 on display at corporate headquarters in Fort Worth, Texas. *-Kenny Rhoads*

Car 32 is seen at Topeka, above, in December 1956, just a few short months after silver paint with black shadow lining has been applied. The car has been modified with SKF roller bearings, air conditioning, fully enclosed roof, communications antennae, and a solid observation platform "railing" has replaced the open original. *-F.O. Kelley photo, John B. McCall collection*

At left, the opposite side of car 32 is seen at Topeka a few years later and the shadow lining has begun to fade. It was then assigned to W. L. Moore, Vice President Operations, out of Chicago. *-John B. McCall collection*

Car 32 is seen here at Chillicothe, Illinois on January 9, 1968, following its final modernization where it received Thermopane windows and abbreviated skirting. The trucks have also been changed out. Its last known assignment, in January 1966, was to the Chief Engineer-System at Chicago. *-Keith Jordan collection*

Business car 32 was renumbered AT&SF 55 on August 19, 1973, at Chicago. The car is now Amtrak compatible and is seen here on No. 4, Southwest Chief at Albuquerque, New Mexico, in the summer of 1975. *-Brian G. Mosely*

As Business car 55, it was sold to Conrail, converted to Track Geometry car CR 31, in October 1983. *-Bill Maier photo, Courtesy Kenny Rhoads*

33
A.T.&S.F.

Built by Pullman in 1923, here we see car 33 at Los Angeles, right, forty years later where it was assigned to the Assistant General Manager, Coast Lines. It has been air conditioned, has received SKF roller bearings and an enclosed observation platform railing and has been shadow-lined. Thermopane windows were applied to most of the heavyweight Business cars during the mid-1960s. In the view below, we see car 33 a bit later with Thermopane windows installed. Attention is called to the consolidation of the three original galley/pantry windows into two at the time the Thermopane windows were applied. Note also that the front window, to left of the diaphragm, originally included in the car as delivered, has been plated over. Car 33 was later sold to a private party, at Chicago in November 1969, and as of this writing is in private car service renamed TAMALPAIS. *-Two photos, John B. McCall*

34
A.T.&S.F.

Business car 34 at Topeka, above, about 1959 has received roller bearings and shadow-line paint, but still retains its front window to the left of the diaphragm. It was at that time assigned to J. B. Noe, Assistant General Manager Eastern Lines at Topeka. *-F.O. Kelley photo, John B. McCall collection*

At Fort Worth, right, a few years later, car 34 has received new trucks and Thermopane windows. *-John B. McCall*

Car 34 was renumbered 56 February 12, 1973, at Ft. Worth, where it is seen in June three years later. It was sold to a Colorado dealer in 1982, who resold it to Conrail for parts in 1984. *-John B. McCall*

Ex Santa Fe Business car 56 stored at Kaiser Steel, Fontana, California, in October 1990. *-Kenny Rhoads*

Car 56 was eventually purchased by Mr. Tom Hoback, president of the Indiana Rail Road, in 2004 who had the car fully restored. Painted in Santa Fe green with gold lettering, car A.T.&S.F. 56 is ready to roll in July 2005. *-Thomas G. Hoback*

Interiors of car 56 as restored by Tom Hoback include, clockwise from left, observation lounge, observation lounge looking toward the rear of the car, and dining room. The original dining room table was re-installed by the Santa Fe during the 1953 rebuild, as well as 1950s-style chairs from another Santa Fe Business car. All chairs were reupholstered with southwestern style fabric typical of those used by the Santa Fe. Flatware and glassware are reproductions from original Santa Fe/Fred Harvey designs. *-All, William R. Carter*

AT&SF 35, built in Pullman Lot No. 4794, poses for its builder's portrait, November 13, 1924. The cars of the "30 series" were virtually identical as released from Pullman. The doorway on the left front was plated to accommodate the interior iceboxes. Structurally, provision was made for possible revamping to include a doorway if later desired, with little rebuilding then being necessary. *-John B. McCall collection*

Track inspection lights were applied to Business 35 by the Pullman Company. This was not standard policy, and the authors have been unable to learn how long this installation lasted in service. Note the short lower section of screen in the end windows. After air conditioning was applied, the screen section was replaced by a full length glass. *-Three photos, Pullman, November 13, 1924, John B. McCall collection*

The interior of the rear observation area on Santa Fe car 35 as fitted at Pullman, included six well padded wicker chairs. An electric fan is provided in an age innocent of air conditioning. The deep, highly polished woods and leaded glass details of the original construction were later replaced by "blonde" paneling in the remodeling efforts of the 1950s. And a small desk, facing to the rear would occupy the space to the left of the rear door, which would be fitted with a speedometer and radio. *-Pullman photo, November 13, 1924, John B. McCall collection*

The observation lounge at the rear of car 35, looking forward with a padded settee against the bulkhead, which could convert into a bed. The passageway at left leads past the bedrooms to the dining room and the pantry and galley beyond. The enclosed metal grill work along the floor of the passage covers the steam heat pipes. *-Pullman photo, November 13, 1924, John B. McCall collection*

The interior of the dining room on Santa Fe car 35 looking toward the galley, at left, and toward the rear of the car, right, as fitted at Pullman. All of the "30 series" cars were similar in design featuring dark polished woods and brass fittings. In view is a small, expandable, dining room table, a set of six dining room chairs, another padded settee which could convert into a bed, and a wood side board fitted with mirrored backs and leaded glass windows. It was in the lockable cupboards and drawers that the china and silver service for the car was kept. An electric fan is also provided. *-Pullman photo, November 13, 1924, John B. McCall collection*

These views of bedrooms C on car No. 35 taken on November 13, 1924, were similar, if not identical to all of the "30 series" cars. Upper left we see the double bed in bedroom facing the rear of the car; facing the front of the car, above, the dresser, folding washstand and hopper in bedroom. Bedroom C was closest to the dining room. The total accommodations contained in car amounted to five double beds and seven single beds, which included the large convertible sofas, or settees, against the bulkheads in both the lounge and the dining room and the facilities for the attendant. *-Pullman Co., John B. McCall collection*

View from the kitchen to the pantry on car 35. The polished metal panel at left was the refrigerator. At right near the window was the sink, covered by a cutting board. Piping carried hot and cold running water to both locations. *-Pullman photo, November 13, 1924, John B. McCall collection*

Detail of wall panel in the dining room on car 35. Polished brass contrasted beautifully with the deep tones of the wood paneling. Attendant call buttons—the lower round fitting—were distributed at convenient points throughout the car. *-Pullman Co., John B. McCall collection*

As one of Santa Fe's remaining heavyweight Business cars in the early 1970s, car 35 was renumbered AT&SF 57 on January 15, 1973, at Topeka. Car 57 is seen here at Amarillo on March 15, 1974. It was later sold and is as of this writing in private car service. *-Joe McMillan*

Car 35 in later years at Topeka, while assigned to R.G. Buchanan, General Manager at Amarillo. The car has the obligatory fully enclosed roof, silver paint with shadow-lining and new trucks. *-John B. McCall collection*

Business car 35, renumbered 57, was sold to G. Merle-Smith, a Denver equipment dealer in May 1982. It was resold to F.L. Mosely in 1983 who renamed it Louisa. As of this writing the former 35 is in private car service in Oregon and named Casper Mountain. *-Courtesy Kenny Rhoads*

36
A.T.&S.F.

Car 36, parked at Lubbock, Texas, in 1960, represented the zenith in heavyweight rebuilding with its completely rounded roof, Thermopane windows, solid observation railing and roller bearing trucks. At the time car 36 was assigned to O. H. Osborn, Vice President & General Manager at Galveston, Texas. *-John B. McCall*

Three years later, in August 1963, the 36 has a fresh set of trucks and has received abbreviated skirting along the side sill. Car 36 is spotted with the MOUNTAINAIR at Topeka. *-John B. McCall*

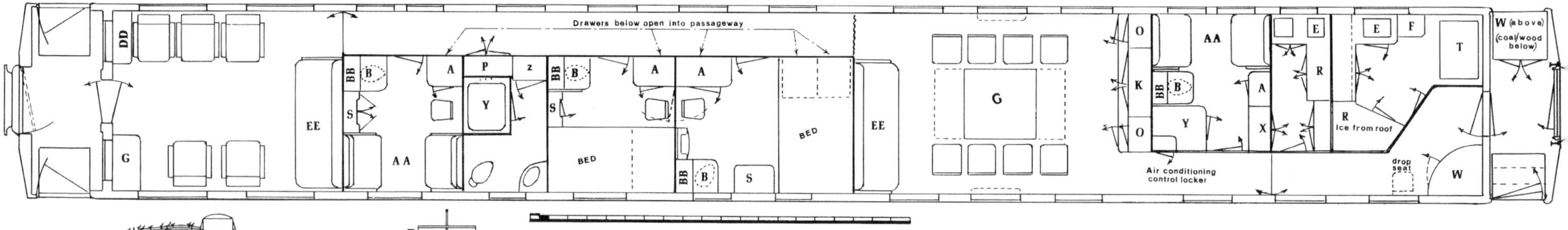

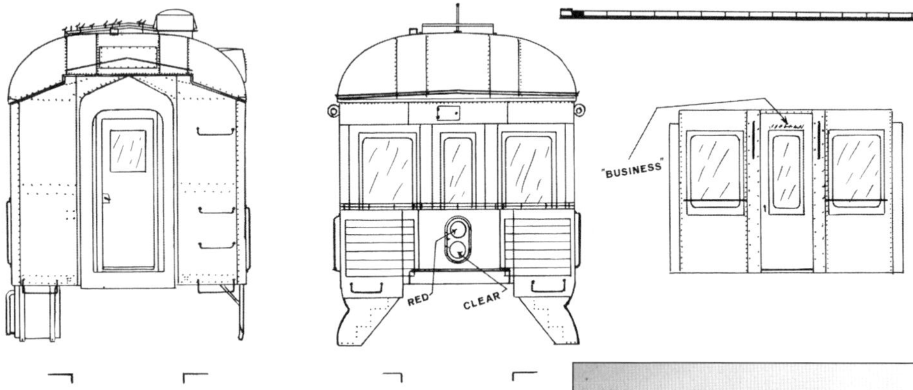

Business Car 36

Drawn as measured

Topeka, Kansas, August 1966

Drawn by Frank M. Ellington

KEY TO DRAWING

A	Wardrobe	P	Locker
B	Hopper	R	Refrigerator
C	Washstand	S	Dresser
D	Ice Box	T	Range
E	Sink	U	Water Heater
F	Drainboard	V	Charcoal Broiler
G	Table	W	Linen/Storage
H	Switchboard	X	Electric Reg. Locker
I	Chime	Y	Shower
J	Speedometer	Z	Shelves Overhead
K	Sideboard	AA	With Upper Berth
L	Annunciator	BB	Folding Washstand
M	Secretary's Desk	CC	Folding Hopper
N	Bookcase	DD	Radio
O	Buffet	EE	Convertible Sofa

At right: Front and rear views of car 36 at Topeka in May 1971.
-Two photos, Frank Ellington

Business car 36 at Topeka in May 1971. Within a year, the car will be retired. Note unusual antenna on the roof over the dining room portion of the car. *-Two photos, Frank Ellington*

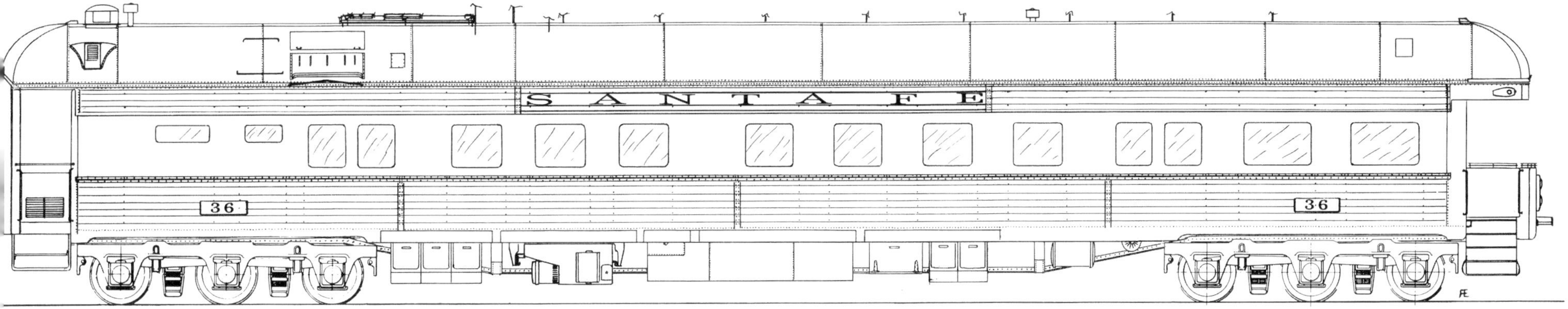

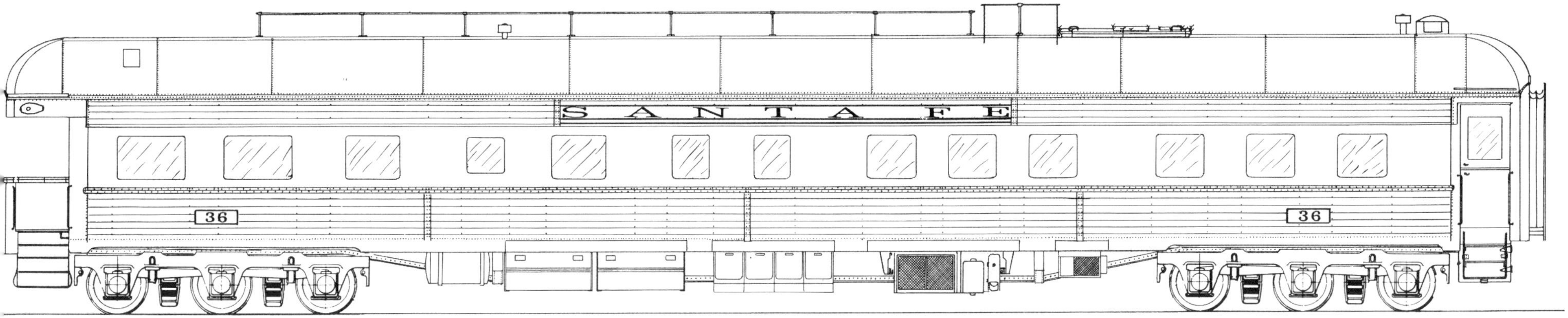

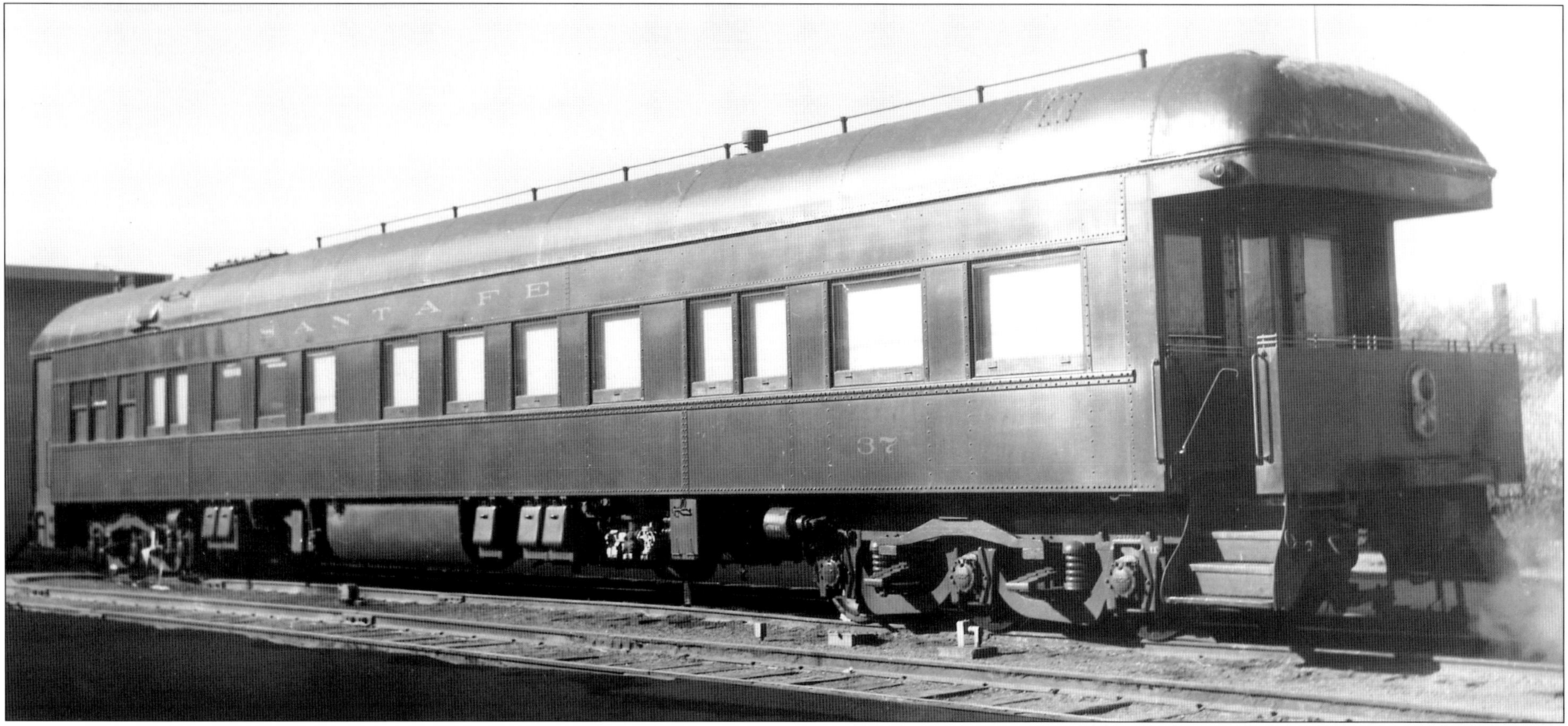

The thirty-three year old Business 37 is still in coach green, but with the modernized roof, SKF roller bearings, and a solid observation platform at Topeka in February 1958. *-F.O. Kelley, John B. McCall collection*

In these Pullman views, opposite, we see the interior of Business car 37 as delivered in December 1925. The larger view, and the view lower right are of the observation lounge. Note the speedometer and train line air gauges mounted near the rear door. The view upper right is of the dining room. Note leaded glass in galley door and attendant call button on wall above table. *-Three photos, Pullman Co., John B. McCall collection*

Car 37 brings up the rear of *The Chief* at Argentine, Kansas, in 1968. *-W.A. Gibson, John B. McCall collection*

PULLMAN

Business car 37 lays over in the train sheds at Fort Worth in October 1967. By now the car was fitted out with a self-contained power plant. *-John B. McCall*

Above and right: Business car 37 on Train No. 26, the Carlsbad District passenger train, at Clovis, New Mexico, March 4, 1968. At the time, the 37 was assigned to the Assistant General Manager of the Gulf Lines *-Two photos, Bob Melbo*

The original assignment of Business 38 was to the General Manager, Coast Lines, at Los Angeles, an assignment it held for its entire service life on the Santa Fe. Car 38, is seen at Los Angeles, right, in the mid-1950s. At this point it has been shadow-lined, and has received SKF roller bearings on its original trucks, as well as the fully-enclosed roof casing. A few years later on a winter's day at Topeka, below, the 38 has been fitted with Thermopane windows and new top equalized trucks. *-Two photos, John B. McCall collection*

Business car 38 was renumbered 58 at Los Angeles on February 16, 1973. Seven months later at Needles, on August 10, 1973, the car retains its shadow-lined sides. Note the addition of the thin skirt along the sill. *-Phil Kauke*

Car 58, running rear forward, brings up the rear of the museum special—transporting all of Santa Fe's historical rolling stock to the California State Railroad Museum—at Grants, New Mexico, on March 1, 1986. The heating and air conditioning on the 30 series cars was originally steam activated and took the steam from the passenger train. After Amtrak, however, the 38 was shopped in early-1973 and self-contained power plants and heating units were installed. *-Bill Childers*

Car 58, seen left in January 1977 at San Bernardino, received a radical new paint scheme in 1974. The shadow striping was done away with and the entire car was painted silver with bright fluorescent orange panels on the front and rear of the car. (Car 55 was also observed sporting the "new look"). Unlike 54 through 57, which were numbered in all four corners just above the trucks, the 58's number was centered on each side. *-Louis A. Marre*

Car 58 still showed in service on the Santa Fe June 11, 1990, but was sold shortly thereafter to regional carrier Arizona & California, which had assumed operations of Santa Fe's former Blythe and Ripley District. Here the car still retains the number 58, but is painted in the Arizona & California colors of green and cream at Parker, Arizona, July 2004. *-John R. Signor*

The last of the "30 series" Business cars, No. 39 was built in Pullman lot No. 6165 and delivered in October 1928. There are several external differences between the 39 and the earlier heavyweight cars delivered by Pullman. Most obvious, was the more modern straight frame equalized trucks under the car. The initials "A.T.&S.F." were no longer applied at the factory and only the number appears at all four corners over the trucks. -*Three photos, Pullman Co., October 11, 1928, John B. McCall collection*

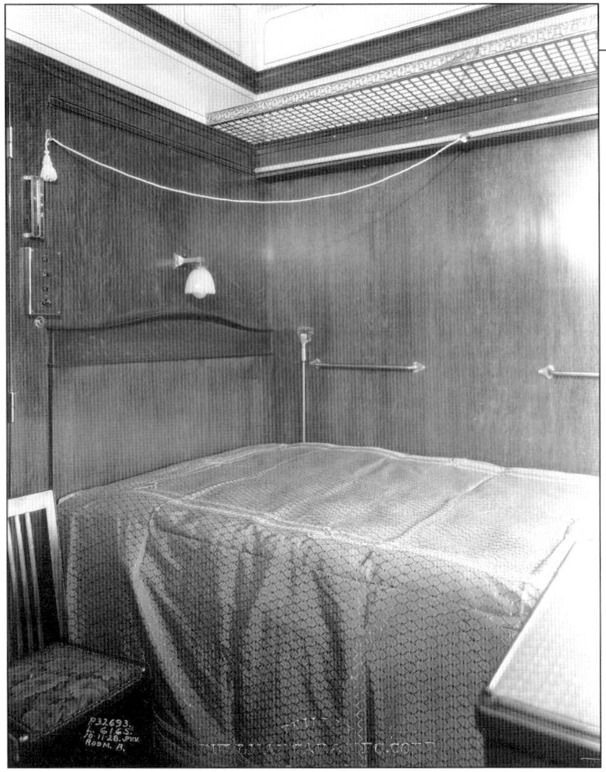

These views of the bedrooms on car No. 39 taken on October 11, 1928, were similar, if not identical to all of the "30 series" cars. Clockwise from lower left we are looking from bedroom A through the adjoining doorway into bedroom B; looking at the bedroom A facing the front of the car; the dresser, folding washstand and hopper in bedroom B facing the rear of the car (note berth stowed at upper left); the passageway between bedroom B and bedroom A; and the dresser, folding washstand and hopper in bedroom C, which was closest to the observation room. The total accommodations contained in the car amounted to five double beds and seven single beds, which included the large convertible sofas, or settees, against the bulkheads in both the lounge and the dining room and the facilities for the attendant. *-All, Pullman Co., John B. McCall collection*

The galley in Business car 39 was compact and utilitarian. In the nearer view can be seen the working counters, sink and cupboards. The opening through the wall facing toward the rear of the car leads to the pantry. In the far view, looking toward the front of the car can be seen the edge of the sink, charcoal-fired range, rear entry door, and linen storage closet. *-Two photos, Pullman Co., John B. McCall collection*

While assigned to the Vice President of Operations out of Chicago, car 39 brings up the rear of a westbound limited at San Bernardino, California, in the early 1940s. The car is essentially in as-delivered condition. *-John B. McCall collection*

Car 39 at Los Angeles after it has been modernized and was assigned to the Chicago pool. Note that Thermopane windows have been applied, except in the pantry/galley area. *-John B. McCall collection*

Car 39, seen here coupled to the MOUNTAINAIR in July 1970 at Topeka, has acquired the thin side skirts and a full compliment of antennae. That's one of the 1952 business car conversions (either 4, 17 or 22) behind car 39, which seems to have a brighter silver carbody and the duller silver roof, although that may be just grime. *-Louis A. Marre, John B. McCall collection*

As the last of the "30 series" cars, the 39 was chosen to be converted to a Track Geometry car during 1973. It was slated to be numbered 5020, but was renumbered 85 prior to being out-shopped on September 4, 1973. For more information on this car see Chapter 8. *-Dean Hale photo, Railroad & Heritage Museum Collection, Temple, Texas*

40
A.T.&S.F.

Santa Fe Business car 40 was built by Pullman in 1927 for Major Max Fleischmann and called Edgewood. It was acquired by the Santa Fe in January 1945 and named Santa Fe. As such it was unique among Santa Fe Business cars at the time for there was no "SANTA FE" on the letterboard, only small "AT&SF" was stenciled on the bottom of the letterboard at both ends of the car. It is seen here at Victorville, California, on the First District of the Los Angeles Divsion on October 14, 1949, while assigned to R.G. Ryden, Vice President, out of Chicago. It was renumbered "40" two months later to make way for the lightweight car of the same name delivered to the Santa Fe by Pullman that same year. -*Chard Walker*

Car 40 is seen at Topeka on September 24, 1950. The Major, evidently, was very sensitive to noise and dust, so the Edgewood was constructed with this in mind. It's sidewalls were stuffed with absorbent material, the floors were covered with four layers of coverings, two layers of wood over steel topped with soundproof cement, with wall-to-wall carpeting over this. There was an instrument panel in the observation compartment which allowed the Major to read train speed, altitude, temperatures both indoor and out, air pressure in the train line and the barometric pressure. There were mirrors permitting occupants to observe the engine and train on curves, and track lights to observe the roadbed at night. -*Ray Hilner collection*

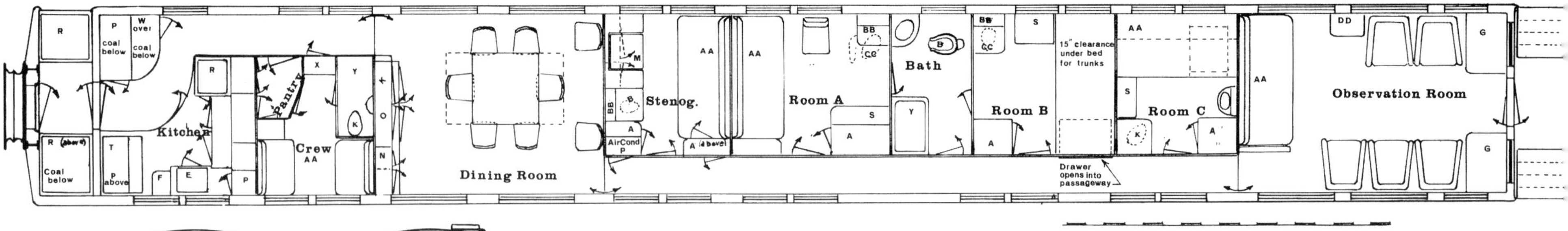

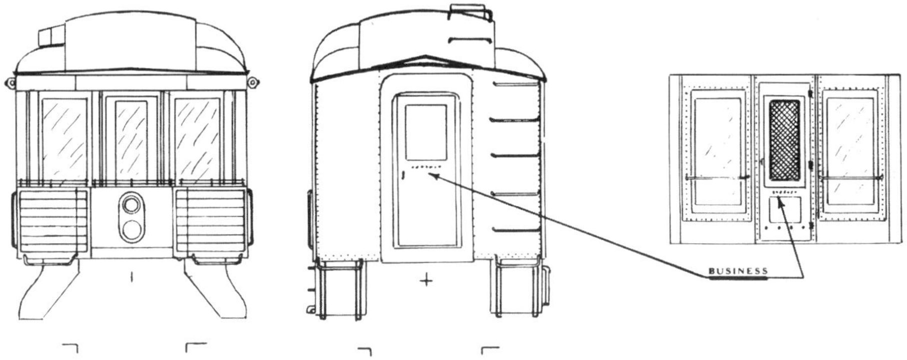

Business Car 40

Drawn as measured
Topeka, Kansas, December 1962
Drawn by Frank M. Ellington

KEY TO DRAWING

A	Wardrobe	P	Locker
B	Hopper	R	Refrigerator
C	Washstand	S	Dresser
D	Ice Box	T	Range
E	Sink	U	Water Heater
F	Drainboard	V	Charcoal Broiler
G	Table	W	Linen/Storage
H	Switchboard	X	Electric Reg. Locker
I	Chime	Y	Shower
J	Speedometer	Z	Shelves Overhead
K	Sideboard	AA	With Upper Berth
L	Annunciator	BB	Folding Washstand
M	Secretary's Desk	CC	Folding Hopper
N	Bookcase	DD	Radio
O	Buffet	EE	Convertible Sofa

A number of Business cars congregate under steam around the Topeka depot in the winter of 1960. While car 40 appears not to be among the cars present, as it was assigned to vice president R.G. Ryden at Chicago at the time, it would have been based out of here. *-John B. McCall collection*

This right side of the 40, at Topeka in August 1966, displays the air conditioning duct on the first air conditioned car in the U.S. The EDGEWOOD was air conditioned during the early 1930s when owned by Major Max Fleischmann. It had already been retired for over three years and was sold for scrap on August 24, 1966 to El Paso Iron & Metal, Vinton, Texas. *-Frank Ellington collection*

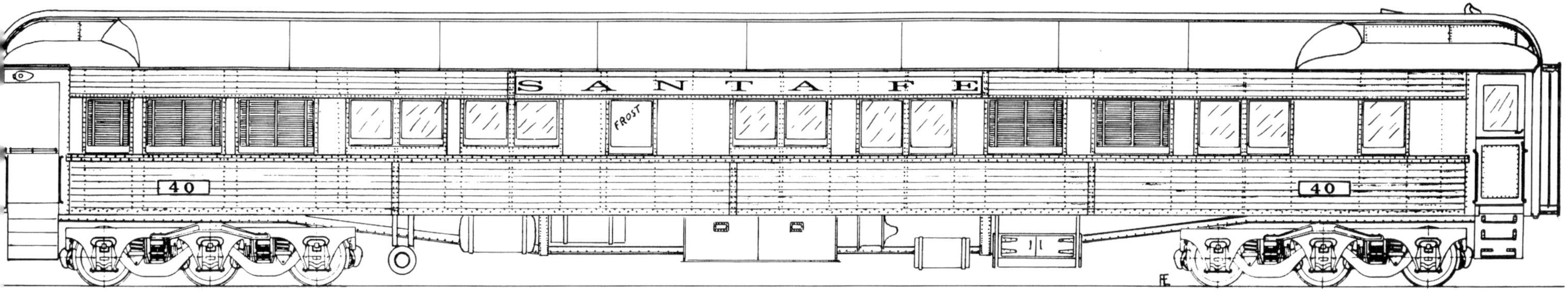

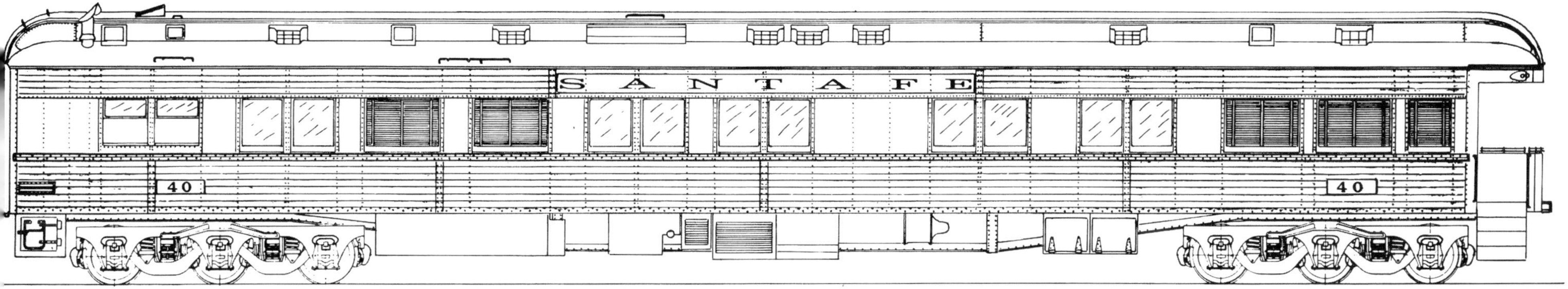

Business car 4—the third car to carry the number on the Santa Fe—was one of three cars rebuilt by Topeka shops in 1952 from 1928-built 3300 series Coaches. During their first few years of service all three Coach rebuilds wore the coach green color scheme. 3rd 4 is seen, left, at Topeka in July 1957. These cars were distinctive for their "C" channel side sills with various cutouts for underbody equipment. *-F.O. Kelley*

Below: During the 1960s, car 4 was assigned to the General Superintendent of Communications and Signaling and displayed a multitude of communications antennae. By now the car had been shadow-lined. Car 4 was renumbered 2nd 21 in July 1967. *-John B. McCall collection*

Car 21, the former 3rd 4, was the most modernized of the three Coach rebuilds as evidenced here by the Thermopane windows at Chicago in July 1968. Even so, the 21 was retired and sold to Grants Supply Co., Tulsa, Oklahoma, in December 1970. *-Frank M. Ellington*

Car 21, the former 3rd 4, later wound up in Mexico as Ferrocarril del Pacifico car 21, Mazatlan. *-Two photos courtesy Mike Martin*

Two identical medium sized Business cars were constructed for the Santa Fe in 1928 in Pullman lot number 6113, cars 2nd 9 and 2nd 10. Note "C" channel side sill. Sleeping accommodations aboard these cars amounted to four double and seven single beds. Here is the 9 at Topeka in 1950. Prior to the application of air conditioning, most of the windows were kept open for ventilation during the warm weather months. *-Charles E. Winters collection*

These window type air conditioners on car 9 at Topeka in 1955 replaced the Waukesha system with the odd ducts indicating the air conditioning applied in December 1953 may have been an experiment. Car 9 was retired in September 1967 and sold for scrap the following May. *-F.O. Kelley*

There is considerable speculation that this car, which has been part of an antique mall complex adjacent to Interstate 5 in Cordelia, California, since the 1970s, is the 2nd 9. *-Kenny Rhoads*

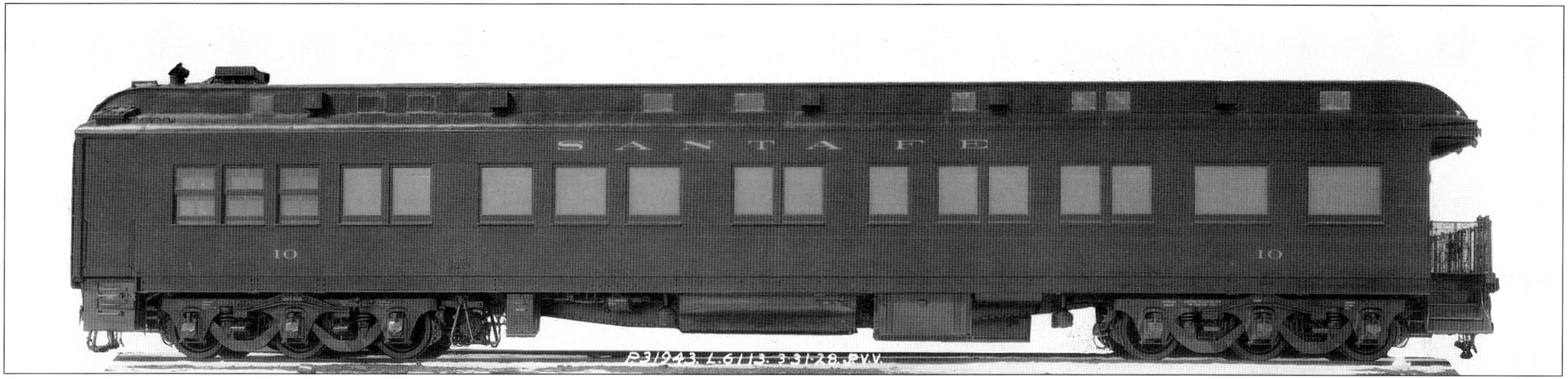

The second car in lot 6113, the 2nd 10, poses for its construction photo at Pullman on March 31,1928. Cars 9 and 10 were numbered to fill vacancies in the rebuilt wood car roster. These cars differed from the "30 series" cars in several ways. One was the absence of an end wall window to the left of the diaphragm on the front of the car. The interiors of these two cars were much more plainly finished, but the intricate observation deck railing was still applied. As it was retired from Business car service to become first a Foreman's car and later a Bunk car, this railing survived intact. *-All, Pullman Co. photos, John B. McCall collection*

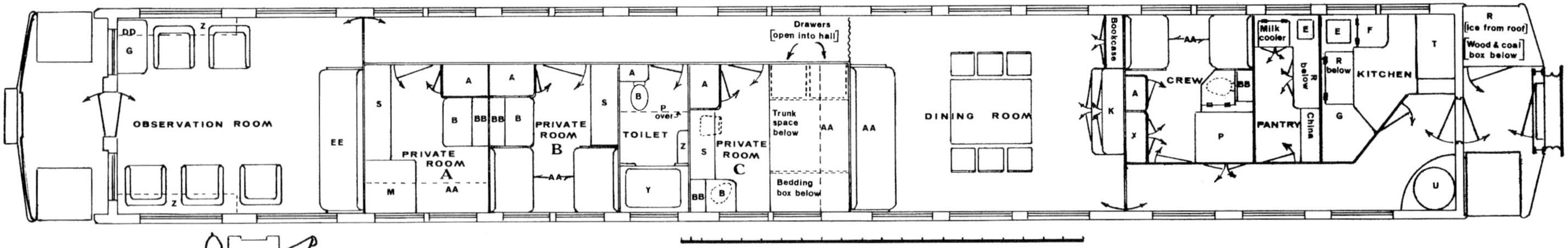

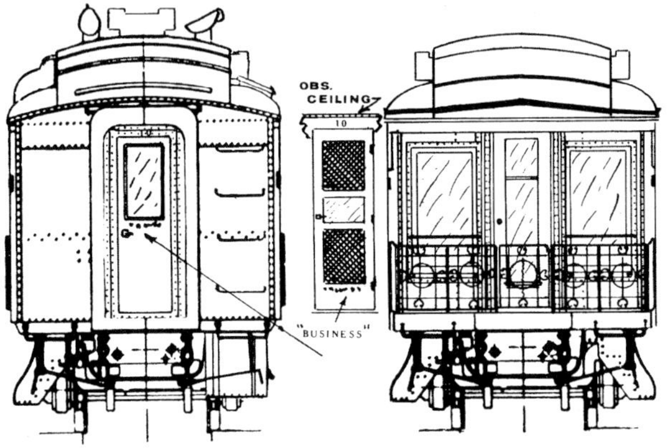

Business Car 10

Drawn as measured

Topeka, Kansas, December 27, 1960

Drawn by Frank M. Ellington

KEY TO DRAWING

A	Wardrobe	P	Locker
B	Hopper	R	Refrigerator
C	Washstand	S	Dresser
D	Ice Box	T	Range
E	Sink	U	Water Heater
F	Drainboard	V	Charcoal Broiler
G	Table	W	Linen/Storage
H	Switchboard	X	Electric Reg. Locker
I	Chime	Y	Shower
J	Speedometer	Z	Shelves Overhead
K	Sideboard	AA	With Upper Berth
L	Annunciator	BB	Folding Washstand
M	Secretary's Desk	CC	Folding Hopper
N	Bookcase	DD	Radio
O	Buffet	EE	Convertible Sofa

Compared to the "30 series" cars, Nos. 9 and 10 were rather spartan on the inside, which with exception of the cabinets, was painted wood. At far left is the observation room, and the dining room. *-Pullman photos, John B. McCall collection*

The interiors of bedroom A in the first two images, and bedroom B at far right, also reflected the plain appearance of cars 9 and 10. *-Pullman photos, John B. McCall collection*

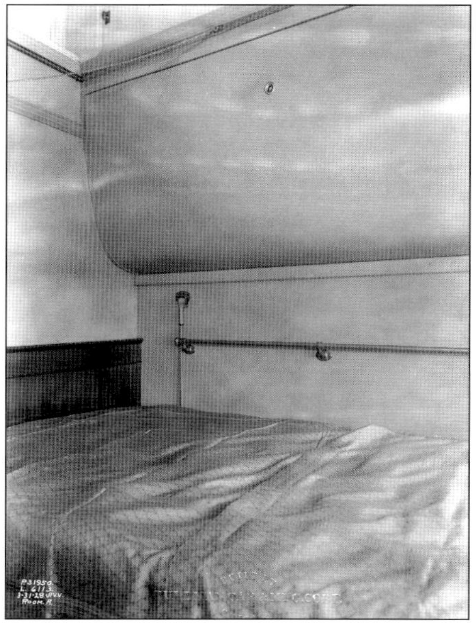

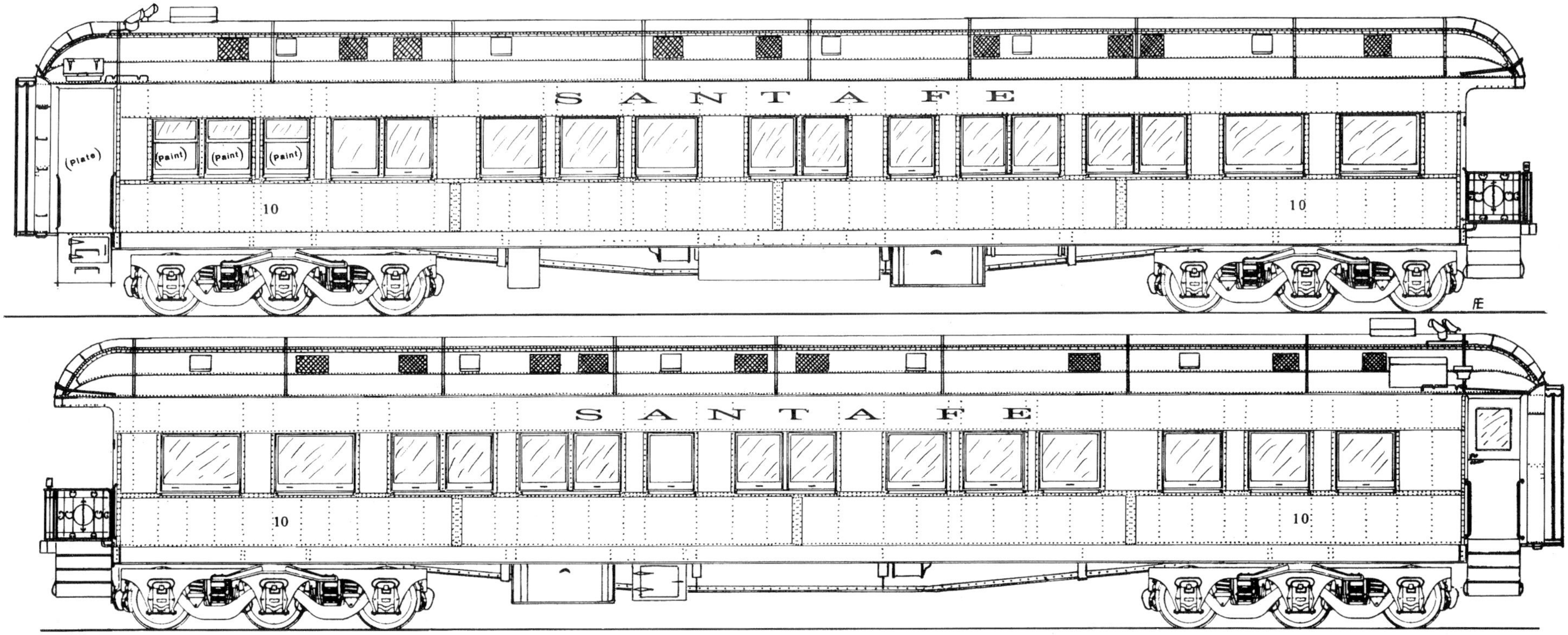

Car 2nd 10 at San Bernardino, March 19, 1941, trailing M-180, possibly on the Second District Local. The 13 year old car has received little modification since leaving Pullman. *-Lewis Harris Photo, John B. McCall collection*

Car 2nd 10 was retired in November 1962, reinstated and converted to Foreman's car 194450 in March 1967. There were only minor changes externally, other than silver paint and the shortened windows in the pantry area. The original observation railing and passenger buffer were left intact. *-Two photos, Railroad & Heritage Museum, Temple, Texas*

Foreman Car 194450 was converted to Living car 203739 in November 1973 at Cleburne, and is seen here at Cadiz, California, right, on June 25, 1986, assigned to ASG Bridge gang. It was later sold and is on private property in Saugus, California, as of this writing. *-Bill Childers*

17
A.T.&S.F.

Car 17 was one of the three cars rebuilt from heavyweight Coaches in 1952. It is seen here at San Angelo, Texas, in December 1962 while assigned to the Assistant General Manager, Amarillo. Note transom windows have been plated over. The three Coach rebuilds were equipped with Waukesha system air conditioning but without exterior roof ducts. *-K.B. King photo, John B. McCall collection*

Car 17 is seen in silver with shadow-lining at Topeka in August 1969 at the end of its service life. *-John B. McCall collection*

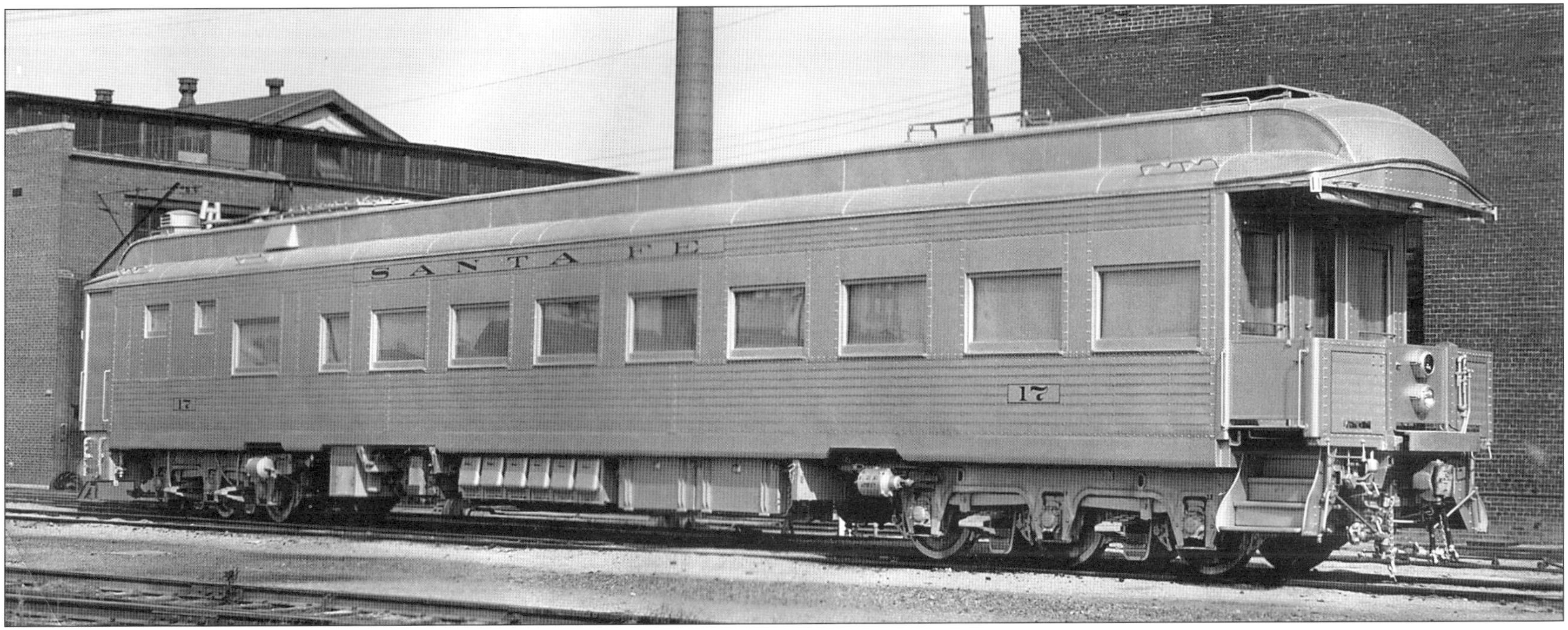

Car 17 has emerged from the paint shop at Topeka in the early 1960s in fresh silver paint with shadow-lining. The car never received Thermopane windows. *-Railroad & Heritage Museum, Temple, Texas*

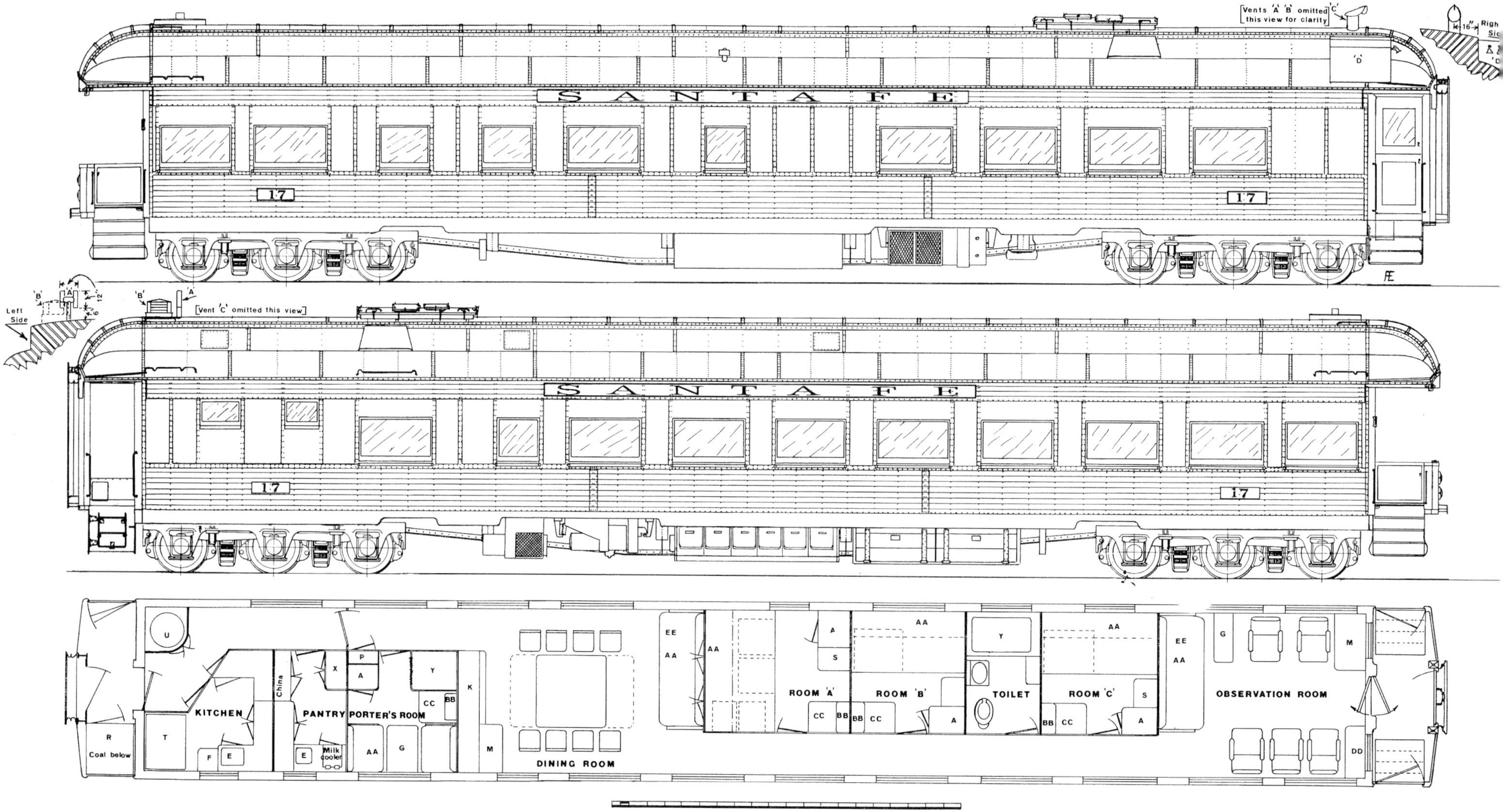

KEY TO DRAWING

A	Wardrobe	I	Chime	R	Refrigerator	Z	Shelves Overhead
B	Hopper	J	Speedometer	S	Dresser	AA	With Upper Berth
C	Washstand	K	Sideboard	T	Range	BB	Folding Washstand
D	Ice Box	L	Annunciator	U	Water Heater	CC	Folding Hopper
E	Sink	M	Secretary's Desk	V	Charcoal Broiler	DD	Radio
F	Drainboard	N	Bookcase	W	Linen/Storage	EE	Convertible Sofa
G	Table	O	Buffet	X	Electric Reg. Locker		
H	Switchboard	P	Locker	Y	Shower		

Business Car 17
Drawn as measured
Amarillo, Texas, August 2, 1967
Drawn by Frank M. Ellington

20
A.T.&S.F.

Originally built by Pullman in 1927 as pool private car Henry Stanley, the 83 ft. 6 in. car was acquired by Santa Fe for a Business car in 1944 and numbered AT&SF 2nd 20. Here it is at San Diego shortly after acquisition dressed in the coach green scheme with its original open work observation platform. *-R.P. Middlebrook photo, Stan Kistler collection*

Nearly twenty years later, Car 20 is seen at Los Angeles Union Passenger Terminal in March 1963. In the intervening period, the 20 has received air conditioning, SKF roller bearings (added in March 1955), silver livery with the shadow line finish, a solid observation platform railing, and additional doors in the front on both sides opening to the interior passageway and to the kitchen. Business car 2nd 20 was sold to Grants Supply Co., of Tulsa, Oklahoma, in January, 1971. *-Morris Abowitz photo, courtesy of Bill Sheehan*

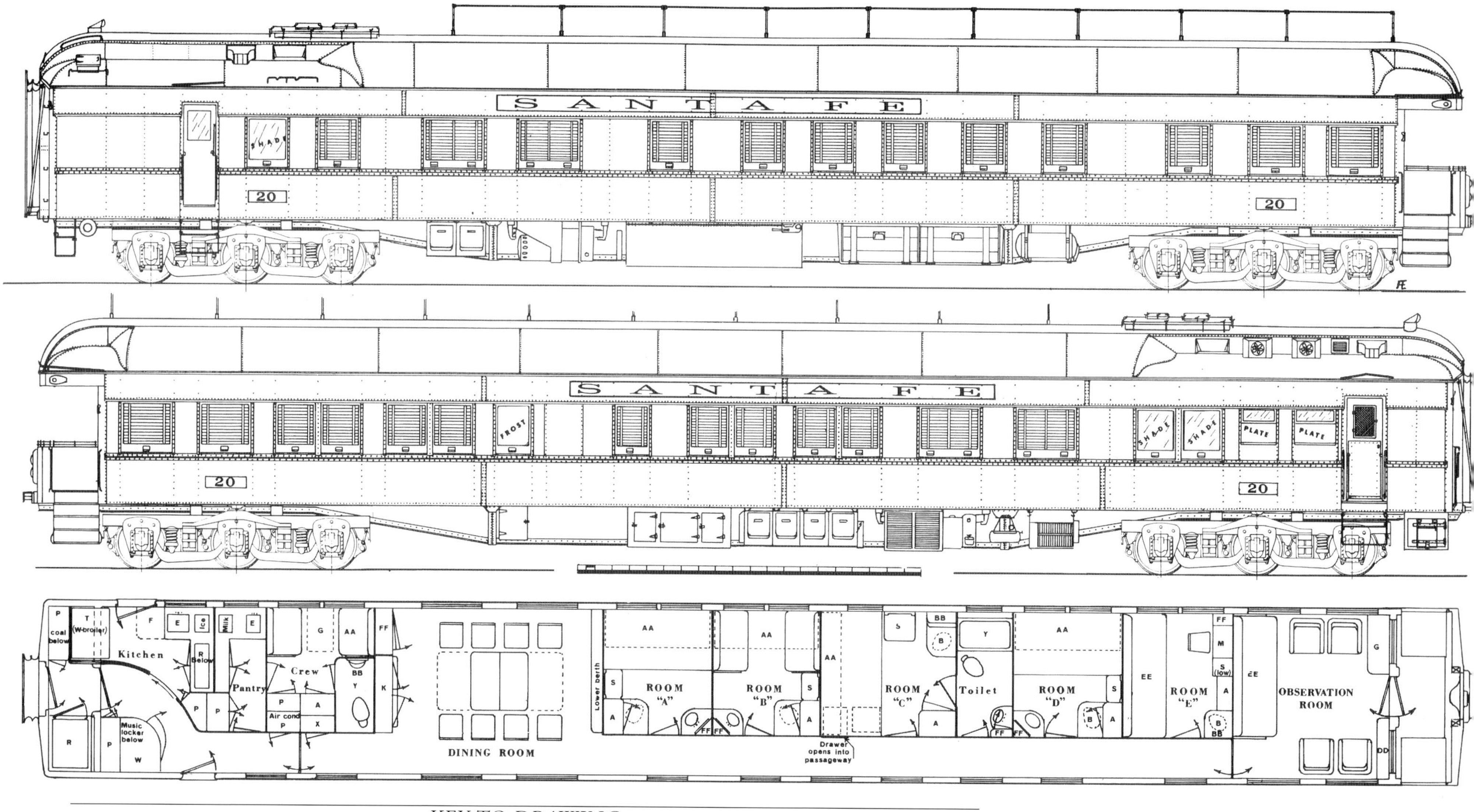

KEY TO DRAWING

A	Wardrobe	I	Chime	R	Refrigerator	Z	Shelves Overhead
B	Hopper	J	Speedometer	S	Dresser	AA	With Upper Berth
C	Washstand	K	Sideboard	T	Range	BB	Folding Washstand
D	Ice Box	L	Annunciator	U	Water Heater	CC	Folding Hopper
E	Sink	M	Secretary's Desk	V	Charcoal Broiler	DD	Radio
F	Drainboard	N	Bookcase	W	Linen/Storage	EE	Convertible Sofa
G	Table	O	Buffet	X	Electric Reg. Locker		
H	Switchboard	P	Locker	Y	Shower		

AT&SF Business Car 20
Drawn as measured
Chicago, July, 1967
Drawn by Frank M. Ellington

Business car 2nd 22 was one of three cars rebuilt by Topeka shops in 1952 from 1928-built 3300 series Coaches. 2nd 22 is seen, left, at Topeka in March 1, 1957, when the car was assigned to the Assistant General Manager at Topeka. Interior furnishings, removed in August 1951 from wood rebuilt Business car 2nd 12 were used in the Coach conversions 4, 17 and 22 in 1952. These cars were distinctive for their "C" channel side sills with various cutouts for underbody equipment. Note that most of the side window transoms have been plated over. *-F.O. Kelley*

Business car 22 in fresh shadow-lined paint at Colorado Springs, right, in November 1965. The opposite side of the 22 appears, above, about 1966 at Topeka. Car 22, like the other two Coach rebuilds, was sold to Grants Supply Co., of Tulsa, Oklahoma, in December 1970. *-Above, John B. McCall collection; right, James C. Seacrest photo, collection Railroad & Heritage Museum, Temple, Texas*

In this interesting view of Santa Fe's LaGrande Station at Los Angeles on September 10, 1933, we find 52-foot steel Superintendent's car 428, coupled to General Manager's car 31, safely tucked in the clear on the private car spur to the north of the depot. In the background, on Santa Fe Ave., Los Angeles Railway streetcars wait departure. *-John B. McCall collection*

5

Steel Superintendent's Business Cars

The Atchison Topeka & Santa Fe was one of the few railroads to possess a fleet of specially-built Business cars for many of their Division Superintendents. This all-steel equipment was placed in service during the mid to late 1920s to replace the old Division Superintendent's cars of wooden framing and underframe construction.

The first four cars, 422 through 425, were delivered by the Pullman Car & Manufacturing Corporation in 1924. With few exceptions, the general design throughout these cars resembled Santa Fe construction standards for steel passenger equipment in general. The underframe consisted of a fish-belly type center sill and channel side sills. Most of the parts comprising the underframe were made interchangeable. Santa Fe construction was worked out so that the entire draft and buffer arrangements with all pertaining parts could be used for any all-steel passenger car. This system of standardizing parts, not only as to castings, kind of specialties or sizes of rolled steel shapes, but as to sizes and shapes of pressed parts and entire construction, provided for an interchangeability, simplifying construction and maintenance of such equipment.

The most interesting feature in the design of these cars was the introduction of a belt rail below the window sill. This rail ran the full length of the car from corner post to corner post and was riveted to all of the side posts. This belt rail signified an advancement in Santa Fe standards of particular interest, as it was introduced for the first time in steel passenger car equipment of the road. The rail gave added strength to the entire side framing, and in this particular construction, served as a support for the outside side sheets, which were applied in long lengths, longitudinally.

When delivered they were assigned to the Oklahoma, Albuquerque, New Mexico, and Galveston divisions. These new all-steel Division Superintendent's Business cars were sheer luxury for their occupants compared to the old 1880-built converted Coaches they previously had occupied.

The interior arrangement of the first four cars included an observation room, one Pullman section with upper berths, one stateroom with stationary bed and one standard upper berth, one dining room, one toilet room, a porter's room, and a kitchen. The car was equipped with double windows made in one part except in the toilet room and the kitchen. The kitchen window was single in two parts, the upper part to drop and the lower part to raise. The upper part was divided into two sections, of which the upper section was provided with a screen and the lower section was provided with glass so that when the sash was dropped the screen covered the opening. The stationary end windows in the observation room were single in one part.

The sofas in the observation room, dining room, and porter's room were arranged with backs to form upper berths. All seats were Pullman standard high back metal frame, with standard seat slides in the Pullman section.

The flooring was made of Flexolith composition with ¼" rubber tiling throughout the car, except the kitchen and the toilet room. The upper floor in the kitchen was covered with 25-lb. copper, 042" thick. The floor in the toilet room was covered with Velvetile floor composition.

The Vapor Car Heating Company's vapor system of steam heat was used. A double coil heater in connection with the hot water heating system was also installed.

Steel Superintendent's car 407 on train 96 at Lubbock in 1959, adds a bit of class to this stub train. *-John B. McCall*

A number of the 400 series Steel Superintendent's cars received this modification in conjunction with the installation of air conditioning in 1950s. *-Lee Berglund, SFRH&MS Collection*

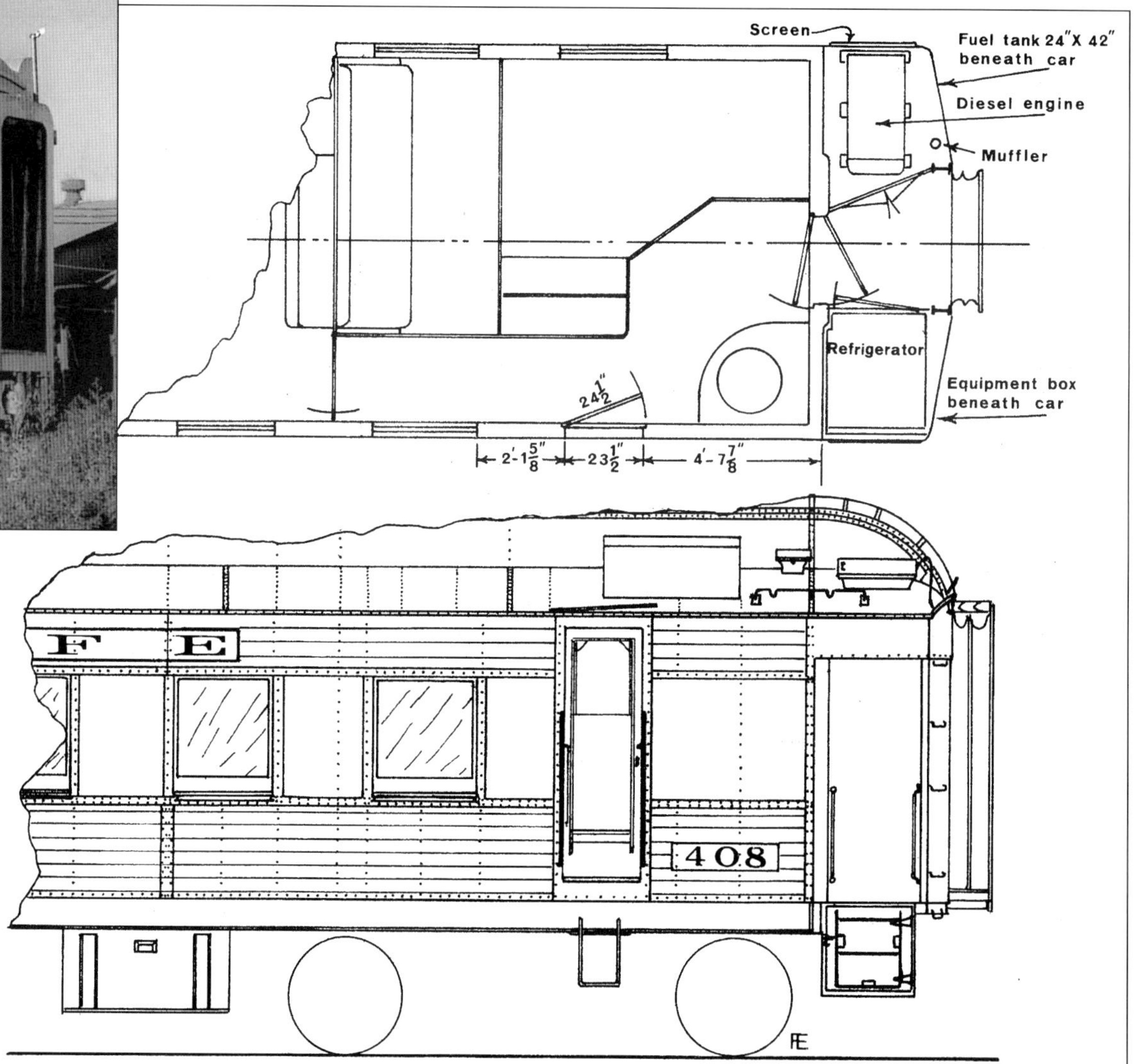

Superintendent's Steel Business Cars Typical Relocation of Crew Vestibule Door

Measured at Barstow, California, 1963

Drawn by Frank M. Ellington

United States Light & Heat Corp. lighting system was used. The axle generator was designed for service on cars having a total load of 46 amperes at 32 volts and would carry a full load at 26 miles per hour car speed. Each car was also equipped with two sets of 16 cells of 13-E lead battery of 240 ampere-hours capacity each, the two sets being connected in parallel.

The cars were equipped with four-wheel trucks, having 37" Santa Fe standard rolled steel wheels and axles having A. A. A. standard 5½" by 10" journals. The truck frame was made of cast steel. The design of the truck followed the Santa Fe standard six-wheel truck very closely. Each truck was equipped with two truck bolsters and a truck center bolster, the latter bolted on to the truck bolster with 16 bolts. By inserting shims between the two bolster units, the proper height of the car was easily maintained at all times.

For a four-wheel truck, the wheel base was a long nine feet. The side frame was 24" shorter than that used on the six-wheel truck; it was in all other respects identical to this side frame. Standard truck parts used on the six-wheel truck were used exclusively in the construction of this truck, and it was therefore apparent that all truck parts not affected by the shorter distance between the wheel centers could

SANTA FE STEEL SUPERINTENDENT'S CARS

Business Car Nos.	*Builder Date Built*	*Lot No.*	*Length & Trucks*	*Notes & Disposition*
AT&SF 2nd 400	Pullman 10/25	4884	52' 4Wh	Converted to 194011 Foreman car 11/23/59 Cleburne; scrap WSCO Anthony, Kansas, 9/17/ 71.
AT&SF 3rd 401	Pullman 11/25	4884	52' 4Wh	Sold, David Wallace, Temple, Texas, 1/24/69; sold to the Texas Transp. Museum, 1969; as of this writing, derelict at Sequin, Texas
AT&SF 2nd 402	Pullman 11/25	4884	52' 4Wh	Sold to Texas Tech., Lubbock, Texas, 11/26/68.
AT&SF 2nd 403	Pullman 11/25	4884	52' 4Wh	Sold to David Wallace, Temple, Texas, 1/24/69; sold to Ute Texas Transportation Museum, 1969; as of this writing under restoration at the Galveston, Texas, Railroad Museum.
AT&SF 2nd 404	Pullman 12/26	4977	52' 4Wh	Sold to David Wallace, Temple, Texas, 1/24/69; sold to the Portland Cement Co., Temple, Texas, 1969; donated by private individual to the Texas Transportation Museum, San Antonio, August 15, 1979; currently under restoration.
AT&SF 2nd 405	Pullman 1/27	4977	52' 4Wh	Sold to the Texas Tank Car Works, 10/25/68; as of this writing, derelict Temple, Texas.
AT&SF 3rd 406	Pullman 12/26	4977	52' 4Wh	Sold to the Texas Tank Car Works, 10/25/68.
AT&SF 2nd 407	Pullman 1/27	4977	52' 4Wh	Sold to David Wallace, Baldwin City, Kansas, 4/3/69; donated to the group restoring the UP North Topeka station.
AT&SF 3rd 408	Pullman 11/29	6250	52' 4Wh	Sold to Barton Salt, Hutchison, Kansas, 10/23/68; preserved and under restoration at the Martin & Frances Lehnis Railroad Museum, Brownwood, Texas.
AT&SF 2nd 409	Pullman 11/29	6250	52' 4Wh	Stored-Calwa, Sold to the Yreka Western A.A., 2/17/66, sold to private individual 1980, at Santa Clarita Ca., (Fillmore & Western) used as residence.
AT&SF 2nd 422	Pullman 5/24	4747	52' 4Wh	Converted to 194010 Foreman car 2/12/60 Cleburne, scrap MPI, 3/29/74.
AT&SF 423	Pullman 5/24	4747	52' 4Wh	Sold to David Wallace, Temple, Texas, 1/24/69; sold to the Portland Cement Co., Temple, Texas, 1969.
AT&SF 424	Pullman 5/24	4747	52' 4Wh	
AT&SF 425	Pullman 6/24	4747	52' 4Wh	Converted to 194009 Foreman Car 12/1/59 Cleburne; retired 7/20/70; scrap WSCO 9/70.
AT&SF 426	Pullman 4/25	4800	52' 4Wh	Sold to David Wallace, Temple, Texas, 1/24/69; sold to the Portland Cement Co., Temple, Texas, 1969.
AT&SF 427	Pullman 4/25	4800	52' 4Wh	Converted to 194012 Foreman car 2/12/60 Cleburne; scrap MPI, 9/29/72.
AT&SF 428	Pullman 4/25	4800	52' 4Wh	Sold to David Wallace, Temple, Texas, 1/24/69; sold to the Portland Cement Co., Temple, Texas, 1969.
AT&SF 429	Pullman 4/25	4800	52' 4Wh	Converted to 194008 Foreman car 10/16/59 Cleburne; retired 1/71; scrap ICS 2/10/71.

be interchangeable with all parts used on the standard six-wheel truck. The long wheel base insured an easy riding car. The trucks were equipped with Simplex clasp brakes.

The Superintendent's Business cars constructed after these first four were furnished with two state rooms, the additional state room replaced the Pullman section with upper berths. The first four cars were shopped in late 1929 and their open section was removed and replaced by the construction of an additional state room.

They were all originally standard green with black roofs, but in later years their roofs were painted aluminum to reflect the heat. Known cars to receive the silver/shadow-line treatment were 401, 402, 405, 406, 408, 424 and 426. Those cars that never received shadow-line had become surplus as the policy to give division superintendents station wagons (instead of Business cars) overlapped the program to paint remaining Business cars shadow-line.

Cars 402, 407 and 424 appear to have been equipped with steam ejector air conditioning. Cars 403-406, 409, 426 and 428 all received package-type air conditioning between 1953 and 1958. The only cars that appear to have been quite extensively upgraded were the 402, and 406, complete with four wheel roller bearing trucks, solid observation platform railing and aluminum painted sides,

STEEL SUPERINTENDENT'S CAR ASSIGNMENTS

Business Car Nos.	*Assignment September 1927*	*February 19, 1954*	*February 8, 1960*	*January 1, 1966*
AT&SF 2nd 400	Oklahoma	H. C. Willis, Oklahoma	-	-
AT&SF 3rd 401	Pecos	T. W. Goolsby, Pecos	T.W. Goolsby, Plains Division, Amarillo	Plains
AT&SF 2nd 402	Valley	F. N. Stuppi, Valley	S. Rogers, Valley Division, Fresno	Valley
AT&SF 2nd 403	Northern	O. D. Crill, Northern	E.E. Baker, Southern Division, Temple	Southern
AT&SF 2nd 404	Eastern	J. F. Fenimore, Eastern	H.D. Fish, Eastern Division, Emporia	Eastern
AT&SF 2nd 405	Western	C. B. Kurtz, Western	C.B. Kurtz, New Mexico Division, Las Vegas	Colorado
AT&SF 3rd 406	Southern Kansas	M. M. Killen, Middle	A.K. Johnson, Los Angeles Division, San Bernardino	Los Angeles
AT&SF 2nd 407	Rio Grande	R. W. Prentice, Panhandle	K.C. May, Slaton Division, Slaton	Stored-Topeka
AT&SF 3rd 408		H. O. Wagner, Slaton	W.A.J. Carter, Pecos Division, Clovis	New Mexico
AT&SF 2nd 409	Missouri	T. J. Anderson, Plains	H.J. Briscoe , Panhandle Division, Wellington	Stored-Calwa
AT&SF 2nd 422	Missouri	J. S. Reed Missouri	-	-
AT&SF 423	Albuquerque	R. M. Champion, Los Angeles	R.G. Alexander, Colo Division, Pueblo	Slaton
AT&SF 424	New Mexico	W. H. Jones, New Mexico	J.P. Spears, Middle Division, Newton	Middle
AT&SF 425	Galveston	J. W. Murphy, Gulf	-	-
AT&SF 426	Illinois	R. J. Yost, Illinois	R.J. Yost, Illinois Division, Shopton	Illinois
AT&SF 427	River	J. P. Spears, Colorado	O.D. Crill, Northern Division, Ft. Worth	-
AT&SF 428	Los Angeles	Sidney Rogers, Albuquerque	-	Northern
AT&SF 429	Beaumont	A. B. Clements, Southern	-	-

with shadow-lining. Car 405 also received roller bearings, but was not further upgraded or shadow-lined.

Cars 400, 422, 425, 427, 429 were retired from the passenger roster in 1960. In 1967, after all Superintendent's Business cars (not previously sent to work service) were retired, the whole surviving fleet was concentrated at Cleburne for disposition.

At the last, green 403 was the Temple car, green 404 was the Emporia car, green 407 (after duty as the Slaton car, replaced temporarily by 423 before it was replaced with a station wagon) knocked around up at Amarillo mainly idle, green 428 was the last Ft. Worth car, one of the first replaced by a station wagon. It, too, was collected at Cleburne in 1967, along with other green cars, and the shadow-lined 400 and 423 cars—all retired. They had been stripped of linens, supplies, china and silver, etc. Car 407, which had been sent from Amarillo to Topeka for a couple of years of storage was already in semi-derelict condition. Its disposition to an exposed location at San Antonio after 1967 completed that job.

As of this writing, several of the 400 series Steel Superintendent's Business cars remain in private or public hands. Cars 403 and 408 are preserved, under protective cover and in the process of being restored at railroad museums in Galveston and San Antonio, respectively. The 409 is being used as a private residence in California. While several other cars are known to have avoided being scrapped, their current condition is either poor or unknown. ✠

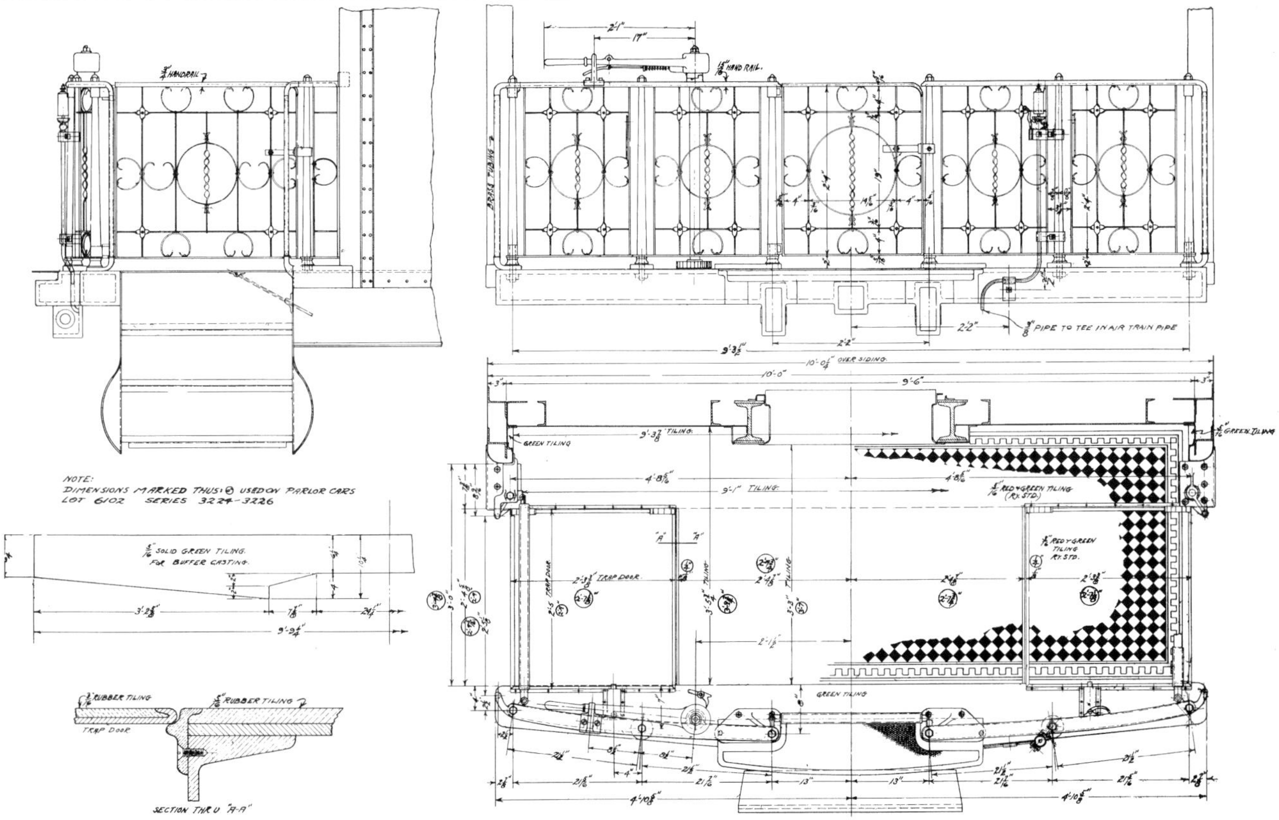

Superintendent's Steel Business Cars
Observation Platform Rail Cars 426-429
Engineer of Car Construction
Chicago, May 14, 1923
Traced by Frank M. Ellington

Opposite: Car 407 brings up the rear of Train 96 leaving Lubbock, Texas, in October 1959 while assigned to K.C. May, Superintendent of the Slaton Division. *-John B. McCall*

400
A.T.&S.F.

AT&SF 2nd 400 is seen at Fort Worth in the 1950s. It was part of the third order of these cars (400-403), delivered in October and November 1925. Apparently assigned to the Oklahoma Division for its career, the 400 was among the first five of the steel Superintendent cars to be converted to Maintenance of Way service. The car was never modernized, air conditioned or shadow-lined. It was converted to Foreman car 194011 at Cleburne on November 23, 1959. *-John B. McCall collection*

These roof shots detail the kitchen end of AT&SF Foreman's car 194011 (ex 400) at Topeka, Kansas, in November 1970. The car was essentially unchanged in this area, except for the electrical conduit running the length of the roof. The two rotating vents were for the heater and range exhaust. The butterfly vent on top was for galley ventilation and the sliding hatch at the corner was for cake ice. The car was sold for scrap September 17, 1971. *-Two photos, Frank Ellington*

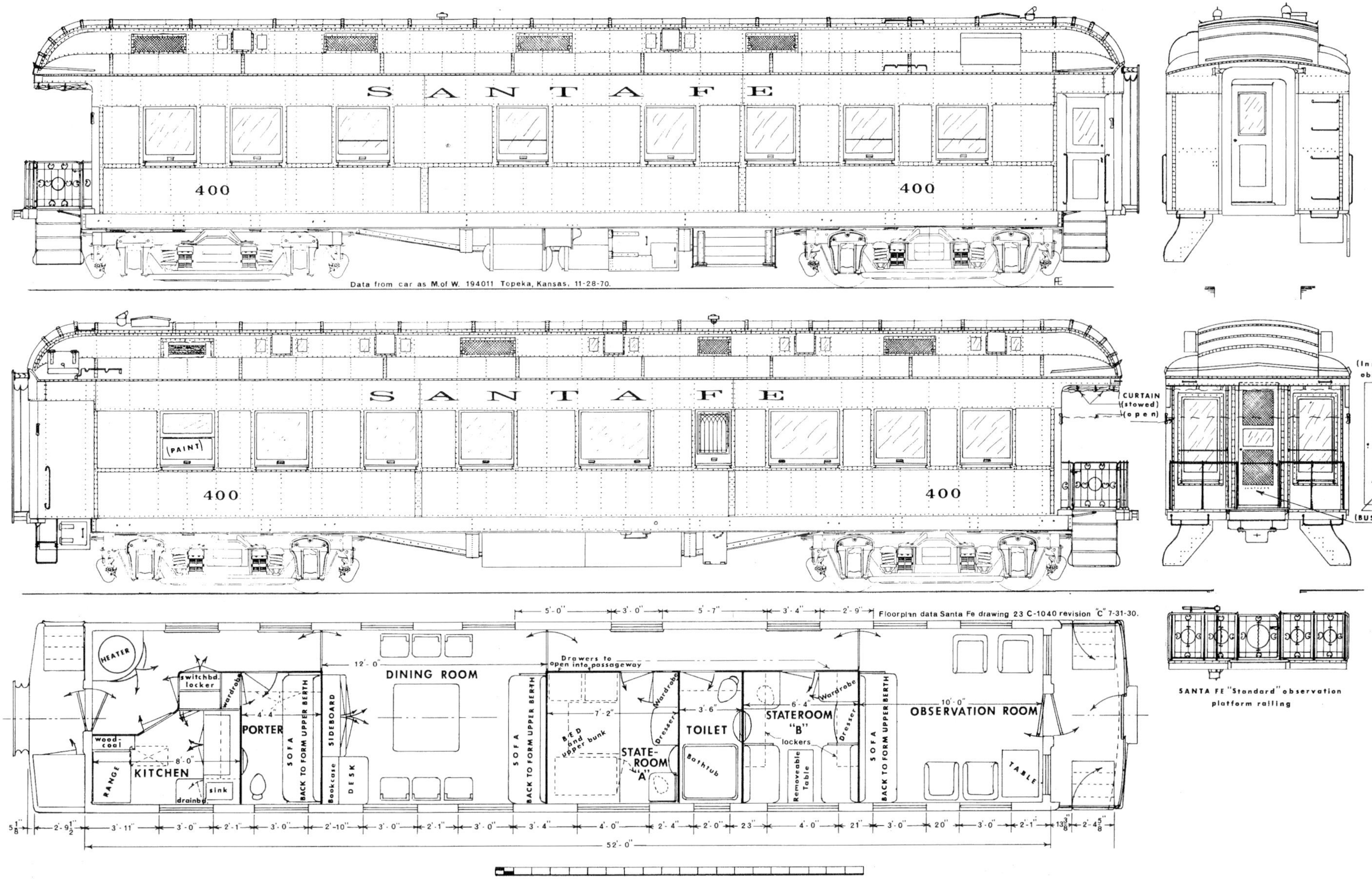

Santa Fe

Superintendent's Business Car 400

Drawn as measured

Topeka, Kansas, November 28, 1970

Drawn by Frank M. Ellington

Steel Superintendent's Car 3rd 401 shows its shadow-line finish to good advantage in this photo at Amarillo, Texas, in November 1967. The car was assigned to the Pecos Division for decades, and the Plains Division at Amarillo in the 1960s. Note the relocated side passageway entry door. The former step area in the right end now accommodated an ice box with roof access for servicing with cake ice. Note also the mirror on the near side window, permitting the occupant to view the train from the rear. The car retained its solid bearings up to retirement and was eventually sold to the Texas Transportation Museum in 1969. *-F.O. Kelley photo, John B. McCall collection*

Details of Car 401 showing the hatch for receiving cake ice, and the ornate railing on the observation end of the car. *-Two photos, John B. McCall*

Steel Superintendent's Business car 2nd 402 shows off the unmistakable Santa Fe construction, with the channel underframe and no belt rail beneath the windows. The car was photographed at Topeka, Kansas, prior to 1950, but was assigned for its entire service life the Valley Division superintendent in California. *-Frank Ellington collection*

The 402 was shadow-lined and received SKF roller bearings in 1959 and is seen here at Richmond, California, the following summer. The 402 and 406 received the Waukesha system air conditioning in 1950 and 1951. While the 402 retained this system, it was removed from the 400 and replaced with the package, or window type, in 1959. The air conditioning duct was on the opposite side of the roof. The 402 and 400 were the only Superintendent's Business cars modified with the solid panel-type observation platform railing. The car was sold to Texas Tech, at Lubbock, Texas, on November 26, 1968. *-Howard W. Ameling collection*

403
A.T.&S.F.

Car 2nd 403, the fourth car delivered in lot 4884 by Pullman in November 1925, is in service at Temple, Texas, December 15, 1962, while assigned to E.E. Baker, Superintendent of the Southern Division. Package type air conditioning using a Buda generating plant was placed in the 403 in June 1955. Note box on the roof over the galley and large intake at the rear of car. Standard coach green, yellow lettering, and a silver roof is the decor. Notice the single Pyle National red signal light. *-Dick Kuelbs*

This page and opposite: Perhaps the Superintendent's car presently in the best condition is the 403, now part of the collection at the Galveston Railroad Museum in Galveston, Texas. Car 3 was retired in 1968 and sold with eight other cars to David M. Wallace of Southwest Rail Services in San Antonio. It was fully restored in 18 months, supervised by Wallace, Ron Bradshaw and Dorthy Hurd. Upon Dorothy's death, her husband donated the car to the Museum in July, 1990. The museum also holds former ATSF 22, which is under restoration. *-All Steve Sandifer*

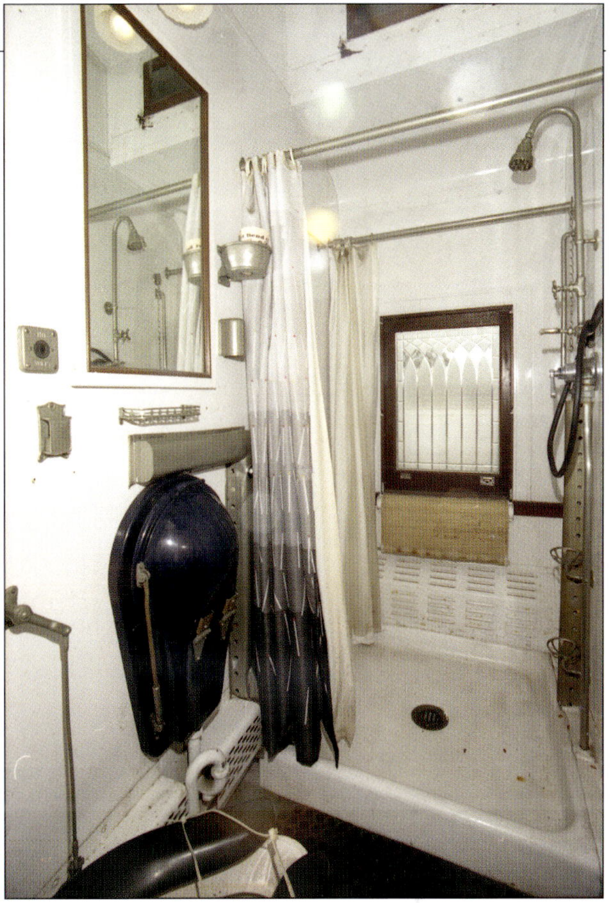

404
A.T.&S.F.

Santa Fe steel Superintendent's car 2nd 404, the first car in Pullman lot 4977 (cars 404-407) delivered in December 1926 and January 1927, is seen in service at Topeka in May 1950. The Superintendent of the Eastern Division, Topeka, Kansas, was assigned car 404 from late 1926 until the late 1960s. *-F.O. Kelley, John B. McCall collection*

The 404 at Topeka, above, in 1954 with a silver roof. The modifications made to the front of the car for package type air conditioning are plainly seen in the view at right at Topeka in November 1964. The curving intake on the roof and grill work on the door and over the original front ventilation screens were part of this work. *-Above, F.O. Kelley; Frank Ellington*

405
A.T.&S.F.

Steel Superintendent's car 2nd 405, looking much as it was delivered, lays over at Dodge City, Kansas, in 1946 while assigned to the Western Division. *-Harold K. Vollrath*

Here we find car 2nd 405 a decade later at La Junta, Colorado. The car had by then received package air conditioning, SKF roller bearings, shadow-line paint, plate radio antenna, and in an unusual application, a rotary gear-type handbrake in place of the lever-type installed as original equipment. The car was sold to the Texas Tank Car Works on October 25, 1968. *-Frank Ellington collection*

Steel Superintendent's car 3rd 406 poses for its builder's photo at Pullman, December 31, 1926. Typical inside and out of all but the last two cars ordered—3rd 408 and 2nd 409—the 406 was 52 feet long and equipped with four-wheel trucks, having 37" Santa Fe standard rolled steel wheels and Simplex clasp brakes. The cast steel truck frame design followed the Santa Fe standard six-wheel truck very closely allowing for interchange of parts. The nine foot wheel base insured an easy riding car. *-John B. McCall collection*

Car 406 in service while assigned to the Middle Division in the 1950s. The 406 received the Waukesha system air conditioning in 1951 and the distinctive air conditioning duct ran the length of the car on this side only. *-Railroad & Heritage Museum, Temple, Texas*

The Waukesha system was removed from the 406 and replaced with the window type in 1959 about the time it was moved west and assigned to A.K. Johnson, Superintendent of the Los Angeles Division at San Bernardino, California. By March 1960, when this photo was taken, the roof had been painted silver as well. *-Bob Melbo*

By 1964, the 406, still at San Bernardino, had been modernized to the extent that it had roller bearings, solid steel platform railing and shadow-lined paint. It was sold to the Texas Tank Car Works on October 25, 1968. *-Jim Lancaster, John R. Signor collection*

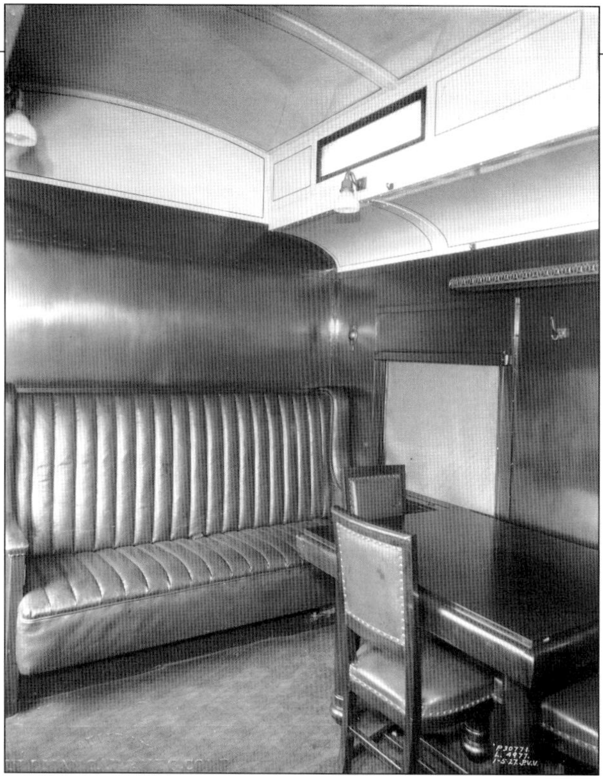

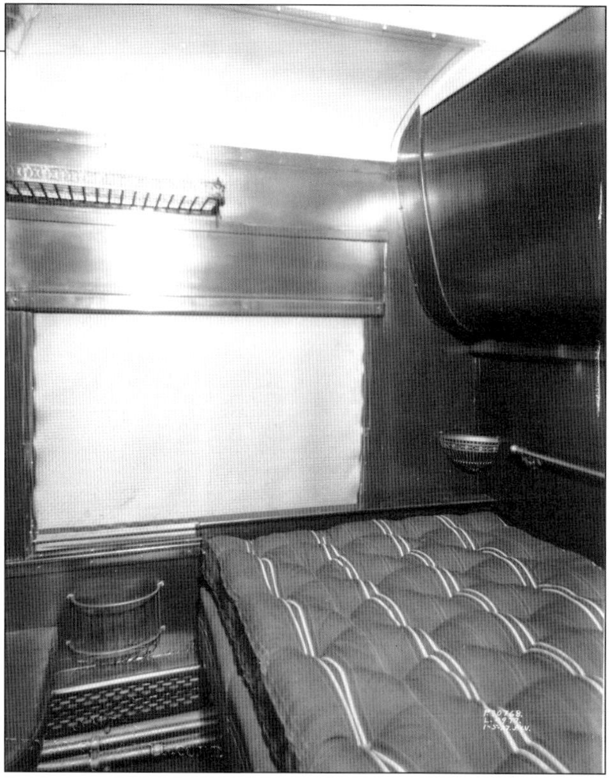

The interior appointments of the 400 series cars were compact and efficient, yet there was touch of comfort for their occupants. These new all-steel Division Superintendent's Business cars were sheer luxury for their occupants compared to the old 1880-built converted Coaches they previously had occupied. The sofas in the observation room, above, dining room, above right, and porter's room, right, were arranged with backs to form upper berths. All seats were Pullman standard high back metal frame, with standard seat slides in the Pullman section. The kitchen and the porter's room were painted in a mahogany color. Pullman photos, January 5, 1927. *-All, John B. McCall collection*

The material for the inside finish was plain, perfectly smooth No. 18 steel which was enameled and baked on before being grained imitation mahogany. The ceiling and the side decks were finished in cream color. The observation room was equipped with a Boyer speed recorder and a Westinghouse air gauge. These instruments were located on the left hand corner side posts. A removable working table 24" by 30" was provided at the right hand corner of the observation room end. The observation room had four wicker chairs with loose cushions. *-Pullman photo, John B. McCall collection*

The dining room of car 406 was typical of the Superintendent's cars. The sideboard, bookcase, dining room table and six chairs, state room dresser and removable tables were all made of Mexican mahogany, as were the doors and sash. All seats and sofas were covered with old Spanish hand-buffed leather of exceptionally good quality. Pullman photo, January 5, 1927. *-John B. McCall collection*

Steel Superintendent's car 2nd 407 was released by Pullman in January 1927. It is seen here about twenty years later at an unknown location while assigned to the Panhandle Division Superintendent. The car was stored at Topeka as early as 1966 and sold to a dealer April 3, 1969 who resold the car to the Steve Sash & Door Co., of San Antonio, Texas. *-John B. McCall collection*

Steel Superintendent's car 407 on train 96 at Lubbock in 1959, bound for Amarillo. The car had probably come in on No. 75, the Houston to Clovis train. *-John B. McCall*

Car 407 was one of the Superintendent's cars which received Waukesha system air conditioning in the late 1940s and early 1950s. Apparently the Waukesha system did not prove adequate for these smaller cars, for the Business cars air conditioned during the mid and late 1950s received the package type air conditioning. In fact, the Waukesha system was replaced on several. This photo of the 407 indicate the Waukesha type roof duct on the right side of the car. Topeka. *-F.O. Kelley*

408
A.T.&S.F.

In service at "home base," ATSF steel Superintendent's car 3rd 408 sits at Clovis, New Mexico, in April 1966 while assigned to W.A.J. Carter, Superintendent of the Pecos Division. The car was shadow-lined after this date. *-Leroy O. King*

Removed from service, we see a now shadow-lined 408 at Cleburne, Texas, in April 1968. The view above illustrates the modifications made to cars with package type air conditioning. A new door was cut into the side of the car opposite the galley. The double-curved box immediately to the right of the side door housed the expansion tanks for the water heating system. The small box to the front of that device is the opening for servicing the coal box, and the large box in the near end is the ice service hatch. At right, the 408 has a gear type hand brake. The car was sold to Barton Salt of Hutchison, Kansas, that October. *-Two photos, Lee Burgland, SFRH&MS Collection*

409
A.T.&S.F.

Steel Superintendent's car 2nd 409 was one of the last two of these type Business cars ordered by the Santa Fe for their division superintendents, 408 being the other. Delivered by Pullman in November 1929 in lot 6250, car 409 differed externally little from the other cars in the series, except for the integral cast pedestal trucks. Pullman photos, November 11, 1929. *-Top, John B. McCall collection; above, Santa Fe Collection, Kansas State Historical Society*

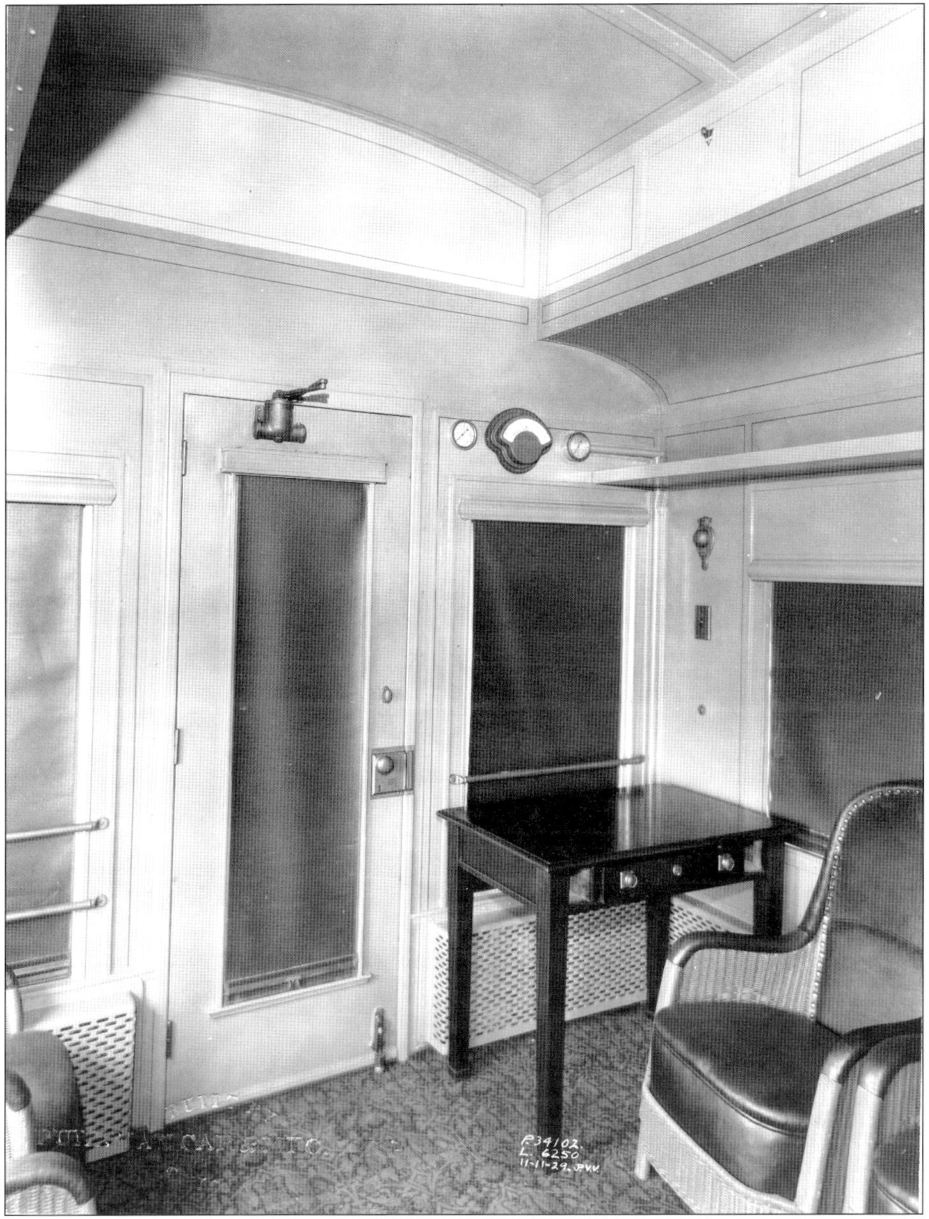

Internally, however, these last two cars were more spartan in appearance, primarily due to the absence of the grained imitation mahogany on the steel walls, window and door trim. The observation room, left, was still equipped with a removable working table and wicker chairs with loose cushions, and a Boyer speed recorder and a Westinghouse air gauge, now on the right hand side. In the dining room, right, the sideboard, bookcase, dining room table and six chairs, state room dresser and removable tables were still all Mexican mahogany, but without the simulated wood paneling on the walls. Pullman photos, November 11, 1929. *-John B. McCall collection*

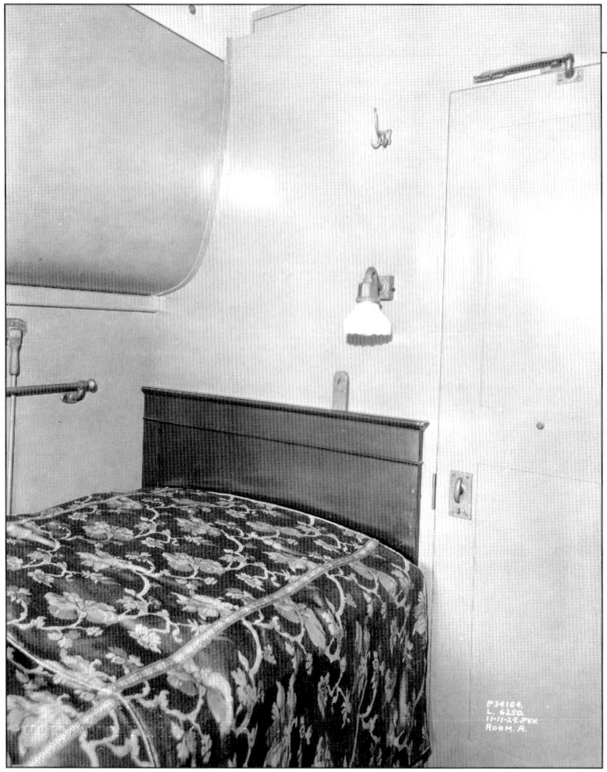
P34104.
L. 6250.
11-11-29. PVK.
ROOM. A.

P34105.
L. 6250.
11-11-29. PVV.
ROOM. A.

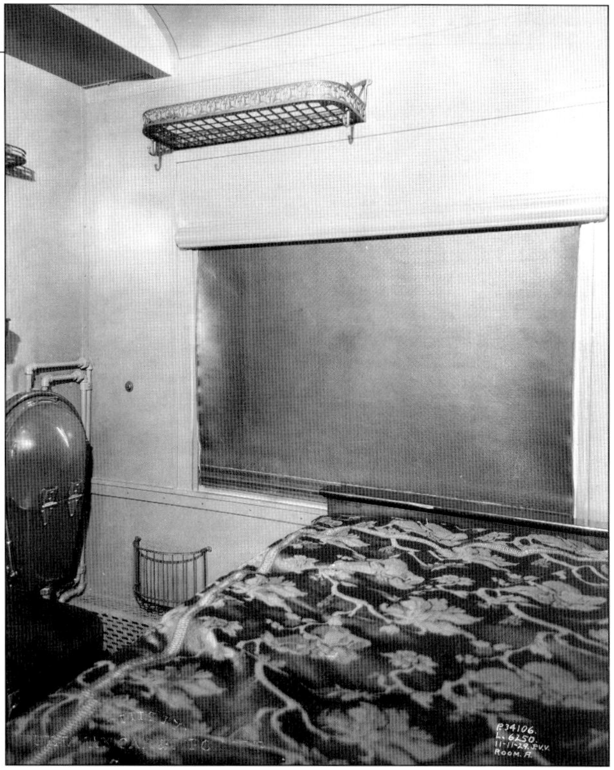
P34106.
L. 6250.
11-11-29. PVV.
ROOM. A.

P34107.
L. 6250
11-11-29. PVV.
ROOM. B.

P34108.
L. 6250.
11-11-29. PVV.
ROOM. B.

P34112.
L. 6250.
11-11-29. PVV.

Originally assigned to the Missouri Division, then the Plains Division in 1954, by January 1960, it was assigned to Panhandle Division Superintendent H.J. Briscoe, above, at Wellington, Kansas. *-H.J. Briscoe collection*

This series of standard photos details the interior of car (2nd) 409. On the opposite page, the top three views are of bedroom (stateroom) A, while the two views lower left opposite are of the secretary's room B. Adjoining the two rooms was a stall shower. The free standing dressers of Mexican mahogany were equipped with a mechanism which prevented the drawers from opening when the car was in motion and the drawers unlocked. In order to open a drawer, one had to rotate the right hand round drawer pull 40 degrees, which retracted a key holding the drawer in place. Above is the steward/cook's room. The device on the wall was a drop down sink. Above right is the range, with the door to the front of the car to the right. Pullman photos, November 11, 1929. *-All, John B. McCall collection*

Car 409 received the package type air conditioning in September 1954 and shows the new side door in this view while stored at Calwa, California in 1965. The car was sold to the Yreka Western Railroad February 17, 1966. It is still in existence today at Santa Clarita California, (Fillmore & Western Railway) and used as a residence. *-Bob Knoll*

422
A.T.&S.F.

Steel Superintendent's car 2nd 422 was the first car of this type built for Santa Fe. Delivered by Pullman in May 1924, it was one of four cars—422-425—delivered in lot 4747 and assigned to the Missouri Division much of its service life. Five of the steel Superintendent's Business cars were converted to Foreman cars for maintenance-of-way crews in 1959 and 1960. Car 424 was converted to Foreman car 194010 at Cleburne, February 12, 1960, and is at Newberry Springs, California, about 1970. It was finally retired and sold for scrap on March 29, 1974. *-Ed Birch collection*

423
A.T.&S.F.

"Shorty" office car ATSF 423 at San Bernardino, California, about 1941 when assigned to Los Angeles Division Superintendent V.H. Wilson, who can be seen on the rear step. The car later went to the Colorado Division and wound up its operating department service on the Slaton Division. *-John L. Whitmeyer, SFRH&MS Collection*

Right: Car 423 at San Angelo, Texas, in 1962 while assigned to the Slaton Division Even in the last years of service, the 423 still wore the coach green with silver roof, Waukesha air conditioning ducts and vents, and the ornate observation railing remained intact. The car was retired and sold to a dealer at Temple, Texas, January 24, 1969. *-K.B. King*

Steel Superintendent's car 424 at Los Angeles in 1933. As part of the first delivery of four cars, the 424 was originally equipped with an open section and one stateroom. During 1929, the open section on these cars was converted to stateroom B. *-John B. McCall collection*

Assigned to the New Mexico Division for nearly 30 years, it was transferred to the Middle Division at Newton by 1960, where it finished out its service life. It is seen here attached to an eastbound train at Topeka, left, about 1957. *-F.O. Kelley*

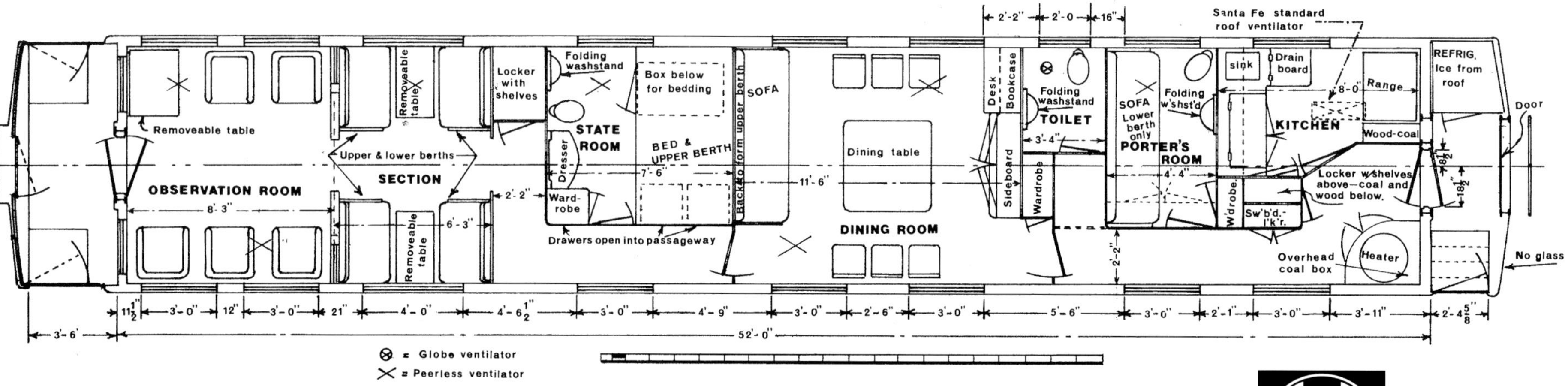

These two drawings, reproduced from badly worn Santa Fe prints serve to demonstrate the basic changes made to the car interiors when the open section area in cars from the first order in 1924 was converted into stateroom "B." In addition to the obvious changes there, the original diminutive 8 ft. 3 in. long observation room was extended to a more comfortable 10 ft. 4 inches. The dining room, porter's room, and kitchen of the car at the front were virtually unchanged. The forward toilet area was not expanded greatly in size, but the removal of the wardrobe and entry way resulted in space savings sufficient to allow installation of a shower in the room. The lower drawing, traced from a 1963 Santa Fe drawing, details the changes, which included two exterior window relocations. *-Pullman photo, Frank Ellington collection*

Superintendent's Business Cars 422-425

Superintendent of Car Construction
Chicago as revised January 26, 1923

Traced by Frank M. Ellington

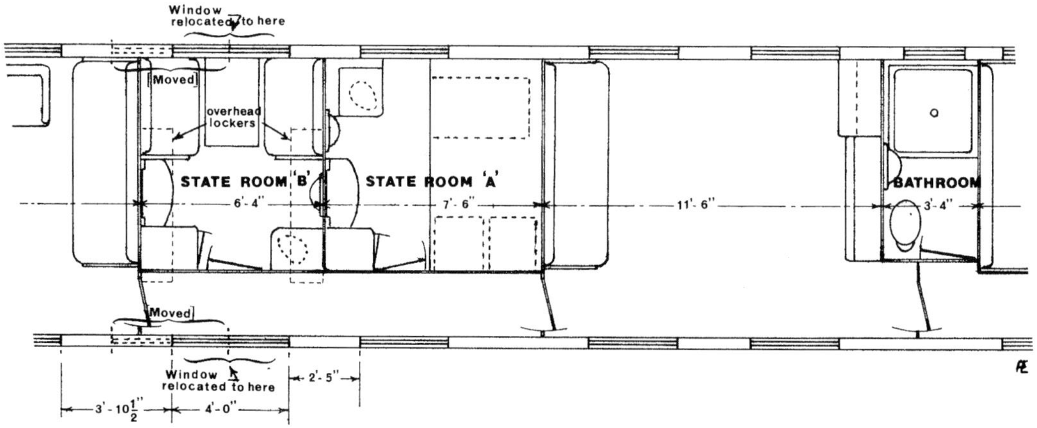

These interiors of car 424, taken by the Santa Fe after the conversion on April 11, 1929, illustrate that, along with the open section portion of the car, all of its simulated Mexican mahogany paint on the steel walls has been painted over. In addition, the furniture, sideboards, cabinets and dressers have all been painted as well leaving little trace of the rich interior originally provided from Pullman, except the hand-buffed leather seat padding. *-Santa Fe photo, John B. McCall collection*

Woodwork, walls and free standing furniture have all been painted in Bedroom A, left. The galley, or kitchen, below, had a hinged board on the counter permitting communication between the galley and the porter's quarters. The window has been dropped, allowing the entry of fresh air—the old style air conditioning method before the mechanical devices of a later period were developed. The bathroom, lower left, showing the shower stall before the curtain has been installed. The large hinged device below the mirror on the left in the photo is the folding lavatory. The stool is in the immediate lower left foreground. Note frosted window installed here. *-Santa Fe photos, John B. McCall collection*

A skilled Santa Fe paint crew has just finished with the job of converting a formerly green-painted car 424 into a shadow-lined silver type, applied to render the car more aesthetically pleasing when operating in consists with fluted-side lightweight passenger equipment. Topeka, Kansas, August 1966. *-Frank Ellington*

Car 424 was assigned to the Middle Division as late as 1966 and is seen here at Cleburne in 1967 *-John B. McCall*

425
A.T.&S.F.

Santa Fe steel Superintendent's car 425, built by Pullman in June 1924, saw service out of Galveston and on the Gulf Division prior to being converted to Foreman car 194009 at Cleburne on December 1, 1959. It is seen at La Junta, Colorado, May 13, 1962. The car was retired July 20, 1970. *-Bill Childers collection*

426
A.T.&S.F.

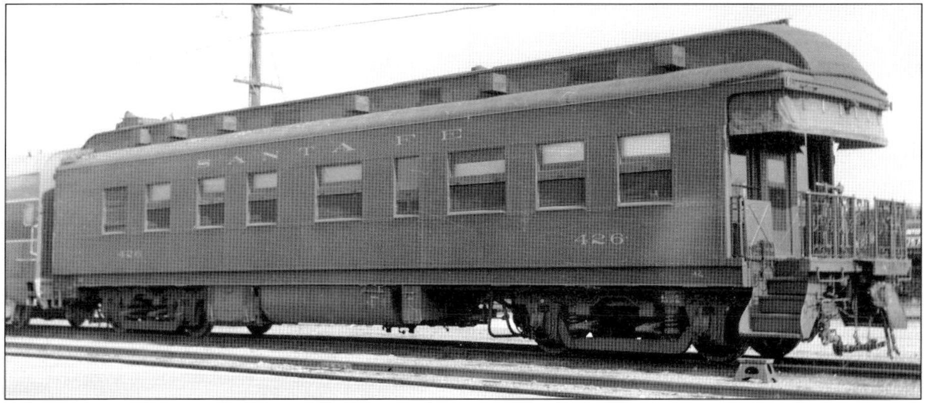

Santa Fe steel Superintendent's car 426, part of the second order of these cars, all delivered in April 1925, was assigned to the Illinois Division for what appears its entire career. In the view, above, it is at Topeka in 1954, while at right, window type air conditioning and a silver roof appear to be the major modifications to the car eleven years later at Shopton (Ft. Madison), Iowa in June 1965, while assigned to R.J. Yost, Superintendent of the Illinois Division at Shopton. *-Above, F.O. Kelley; right Frank Ellington*

In 1967, after all Superintendent's Business cars (not previously sent to work service) were retired, the whole surviving fleet was concentrated at Cleburne for disposition. Here we see the 426, now shadow-lined, awaiting its fate. Note the vestibule step was not removed when the doorway was sealed to accommodate the refrigerator. The car was sold to a dealer the following year. *-John B. McCall*

427
A.T.&S.F.

Steel Superintendent's Business car AT&SF 427 was built by Pullman in April 1925 in lot 4800. Its last Operating Department assignment was to O.D. Crill, Superintendent of the Northern Division at Fort Worth before being converted to 194012, Foreman car, on February 12, 1960, at Cleburne. It was sold for scrap September 29, 1972. As of this writing, no photos of this car have come to our notice.

428
A.T.&S.F.

Superintendent's Business car AT&SF 428, also built by Pullman in 1925, was last assigned to the Northern Division, before being sold to dealer David Wallace, at Temple, Texas, January 24, 1969, who resold it to the Portland Cement Co. at Temple. *-Left, SFRH&MS Collection; above, John B. McCall*

Car 428, being pulled by GP 9727 at an unknown location in Texas about 1958. Note the portable lights attached to the rear observation railing, as compared to the more permanent fixture in the above two views. *-John B. McCall collection*

429
A.T.&S.F.

Photos of Steel Superintendent's car AT&SF 429 in Operating Department service are hard to come by. The car was converted to 194008, Foreman car, at Cleburne October 16, 1959, and is seen here spotted on a siding at Lamy, New Mexico, on June 2, 1960. It was retired and sold for scrap February 10, 1971. *-John B. McCall*

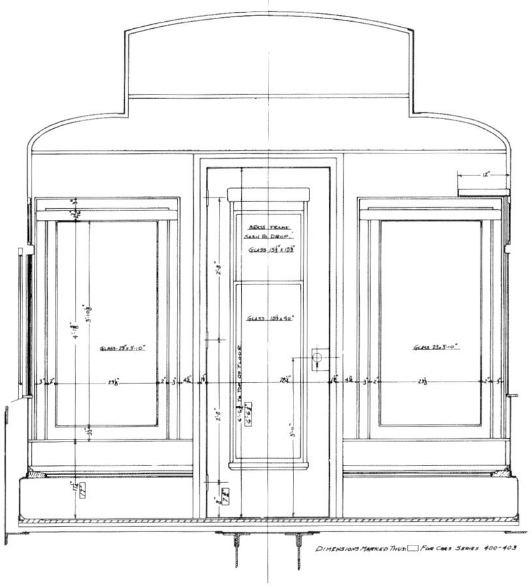

End Finish Observation Room

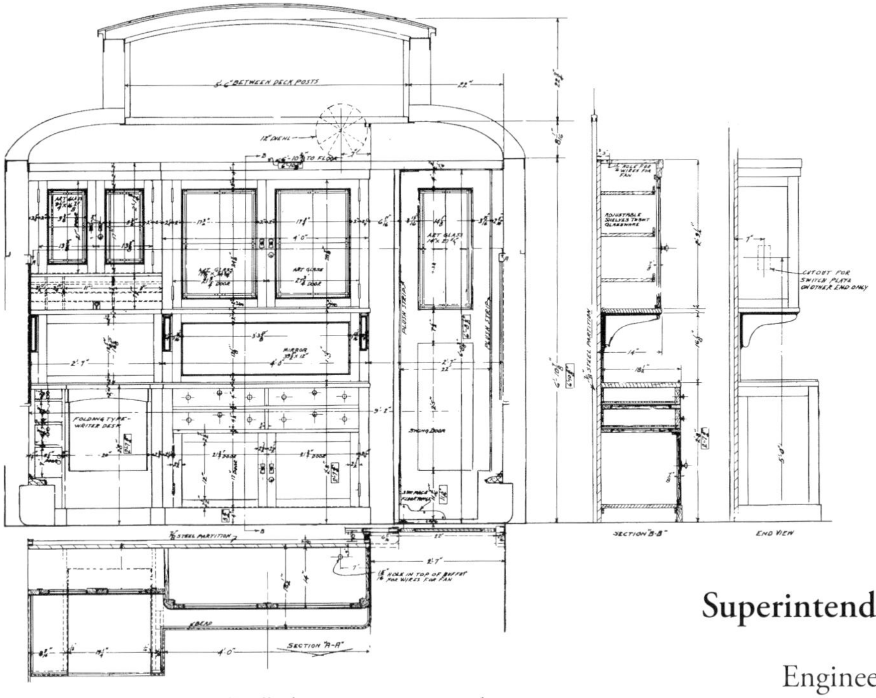

Partition "G" dining Room Side

Superintendent's Steel Business Cars
426-429
Engineer of Car Construction
Chicago, May 14, 1923
Traced by Frank M. Ellington

As viewed from the mezzanine of gloomy Dearborn station in Chicago in June 1968, Santa Fe lightweight Business car ATCHISON, built by Pullman in 1949, is occupied and attached to No. 17, the westbound *Super Chief*. No. 17 is due to depart track 5 at 6:30 p.m., and the executive, after partaking of a light dinner will get some more work done before retiring after crossing into Missouri later in the evening. *-Richmond S. Bates*

6
Lightweight Business Cars

Santa Fe's first lightweight Business car was the 2nd SANTA FE, ordered new from Pullman and delivered in 1949. It was 85 feet long and had four-wheel trucks. When delivered, the first car to carry that name, the 1927-built Pullman private car, which Santa Fe acquired in January 1945, reverted to its number only, ATSF 40.

In 1957, the Santa Fe Railway received two new lightweight stainless steel Business cars named the TOPEKA and the 3rd SANTA FE. These two cars were among the very last business cars built new by any builder for any American railroad. (There have been several business cars built since that time for other railroads; but, these have usually been rebuilds from other cars, not built new by a manufacturer.)

Santa Fe ordered these two cars in January, 1956, and delivery from the Budd Company was made in August 1957. The Budd job number and lot number for these cars was 9664-186. In Budd's records, the SANTA FE was built as Budd car number 3891, and the TOPEKA was built as Budd car number 3892. As built, the cars were 87 feet 6 inches long, and were equipped with six-wheel outside spring hanger trucks, with Budd disc brakes.

The roofs of the cars were decorated with a wide assortment of antennae, including radio-telephone, television, broadcast receiver, and VHF antennae. The interior arrangements included kitchen, pantry, and dining areas. The crew dormitory space next to the kitchen and pantry areas included shower space, a feature seldom seen in any normal sleeping cars, or other Business cars.

There were four bedrooms. Bedroom A had a lower berth and a folding bed; bedrooms B, C. and D each had lower and upper berths. In the rear observation lounge area, there was also a convertible sofa. Beside the normal bathroom facilities in each bedroom, there were also two showers in the car for the bedroom occupants.

The cars were electrician's nightmares because they had four separate electrical systems. The primary system was 220 volt AC; the secondary system was a 110 volt AC. There were also 32 volt and 12 volt DC electrical systems. The cars were equipped with two Detroit Diesel engines, each connected to a 35 kilowatt, 220 volt self-regulated alternator. The cars also had four receptacles, two on each side, for standby stationary power.

As the new SANTA FE and TOPEKA cars were delivered from Budd in August 1957, the 1949 Pullman-built lightweight SANTA FE, was renamed the ATCHISON.

As delivered, the cars had very little in the way of lettering. The car names were centered on the sides of the cars but the railroad name did not appear on the letterboard. Instead, there was simply a small "ATSF" in the letterboard space near the front of the car. The TOPEKA and SANTA FE were assigned numbers 5002 and 5003 respectively, for computer and accounting purposes, but these numbers never were applied to the cars. But, in late 1973, these cars, along with Santa Fe's other named Business and Director's cars, were assigned car numbers, which were applied to the cars themselves. The SANTA FE was numbered 50 on September 30,1973, at Chicago; and the TOPEKA was numbered 51 on September 27, 1973, at the shops in Topeka and the ATCHISON number 50, on August 8, 1973. These numbers were added to the exteriors of the cars in small numerals below the "ATSF" located in the letterboard space near the

The 1949 Pullman-built Business car ATCHISON at Chicago about 1960. This was originally the 2nd SANTA FE. *-Joe McMillan*

Security Inserts

By April 1979, all three lightweight open platform cars Atchison, Topeka and Santa Fe, had been equipped with removable stainless steel inserts and temporary diaphragms (some times referred to as "security inserts") fabricated at Santa Fe's Topeka Shop. Fitted to the observation platform end of the car, installing the inserts resulted in a fully enclosed car with conventional dutch doors suitable for mid-train use under all weather conditions. It is unclear how the inserts were attached, and where they were stored when not in use, but photographs of these cars exist with, and without, the inserts in place after April 1979. Apparently, Business car 1869 (former 54, originally 31) was the only heavyweight known to have been so equipped.

Business car Atchison, fitted with stainless steel insert in closed position, at Topeka, Kansas, April 15, 1979. *-Frank Ellington*

Right: Details of Business car Topeka fitted with the insert in open and closed position, at Topeka, Kansas, on April 15, 1979. The installation on the Santa Fe was identical. Notice how the insert nests inside the platform handrails.
-Four photos, Frank Ellington

front of the cars.

Santa Fe purchased its only secondhand lightweight business car from United States Steel Corp. in August 1962. The car, while owned by U.S. Steel was named LAUREL RIDGE, and it emerged from Topeka shops renamed AT&SF MOUNTAINAIR in November 1962. This was an 85' car with an enclosed observation room. Originally 5004, it was renumbered car 53. It was renamed JOHN S. REED in 1990 to honor the well-respected former president and later director of the Santa Fe.

During 1990, the Budd six-wheel cars, TOPEKA and SANTA FE, were re-equipped with four wheel trucks, presumably for a shorter wheel base, which would allow the cars to navigate industrial trackage. At the same time, the cars were fitted with wood paneling to make the rather Spartan interiors more appealing. As the story has it, on one of the first trips out of the shops, the cars rode so rough that some of the wood paneling was pulled loose.

These four cars saw extensive usage on Santa Fe's business and director's specials over the years, ranging over most of the Santa Fe Railway system. The car SANTA FE was normally designated as the President's car on these specials. When not in use, they were stored with Santa Fe's other remaining passenger equipment inside the Topeka passenger shop buildings where they were secure from vandals.

All four of these cars survive as of this writing as part of BNSF's active Business car fleet.

SANTA FE LIGHTWEIGHT BUSINESS CARS

Business Car Name	*Builder Date Built*	*Length*	*Trucks*	*Notes & Disposition*
SANTA FE 3rd	Budd 1957	87'6"	6wh	Assigned number 5003*; renumbered 50, SANTA FE, 9/30/73 at Chicago; to BNSF 7, SANTA FE.
TOPEKA	Budd 1957	87'6"	6wh	Assigned number 5002*; renumbered 51, TOPEKA, 9/27/73 at Topeka; to BNSF 6, TOPEKA.
ATCHISON	Pullman 1949	85'	4wh	Delivered as 2nd SANTA FE; renamed ATCHISON 8/57; assigned number 5001*; renumbered 52, ATCHISON, 8/8/73 at Chicago; to BNSF 5, ATCHISON; to BNSF 88, Track Measurement support, no date, (retains name ATCHISON).
MOUNTAINAIR	Pullman 1949	85'	4wh	Originally LAUREL RIDGE, acquired by Santa Fe 11/62, renamed MOUNTAINAIR; assigned number 5004*; renumbered 53, MOUNTAINAIR, 8/21/73 at Chicago; renamed JOHN S. REED, 1990; to BNSF 8, JOHN S. REED.

* *The numbers 5001 through 5004 were for computing purposes only; the exterior of the cars were not so numbered.*

Business car TOPEKA has come to a stop after becoming detached from train No. 4, the *Southwest Chief*, east of Springer, New Mexico, in March 1980. No. 4 was clipping along at a pretty good pace when the incident occurred. Later, arriving at Raton, there was a noticeable clack of flat wheel spots on the car. It was a bad day no doubt for the roadmaster. *-Orville Pratt*

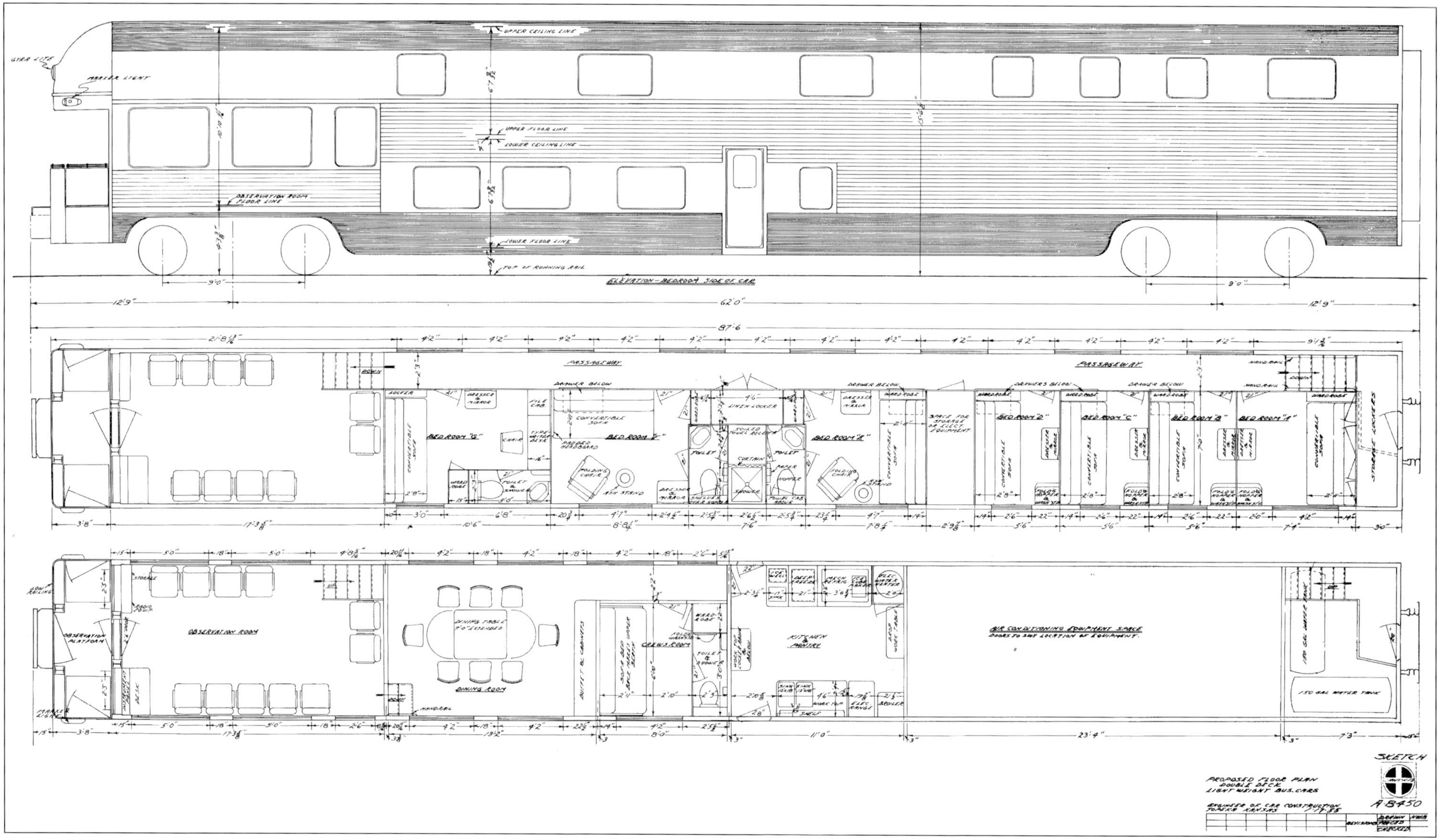

What Might Have Been. In January 1955, as Santa Fe was investing heavily in Hi-Level equipment and experimenting with the concept for multiple applications on the *El Capitan* and later *San Francisco Chief*, a draftsman in the Engineer of Car Construction office at Topeka was asked to rough out a new business car design that might agree in line with the increasingly popular Hi-Level equipment. On this and the following page are the results. Of course, many hundreds of detailed drawings were necessary after acceptance by the authorities of the general outline work represented by these two general arrangement types, but it is interesting to speculate as to what form the finished product might have taken in its emergence from the car builders bays. Note that in both of the drawings the "split-level" concept did not carry over into the observation room area, that portion being assigned a spaciousness best described as the "cathedral ceiling" concept, with almost an 11-foot clear area between the "deck" and the ceiling. Interestingly, too, Santa Fe seemed destined to retain the open observation platform design as a convenience to its occupying officials. *-Richard E. Scholz collection*

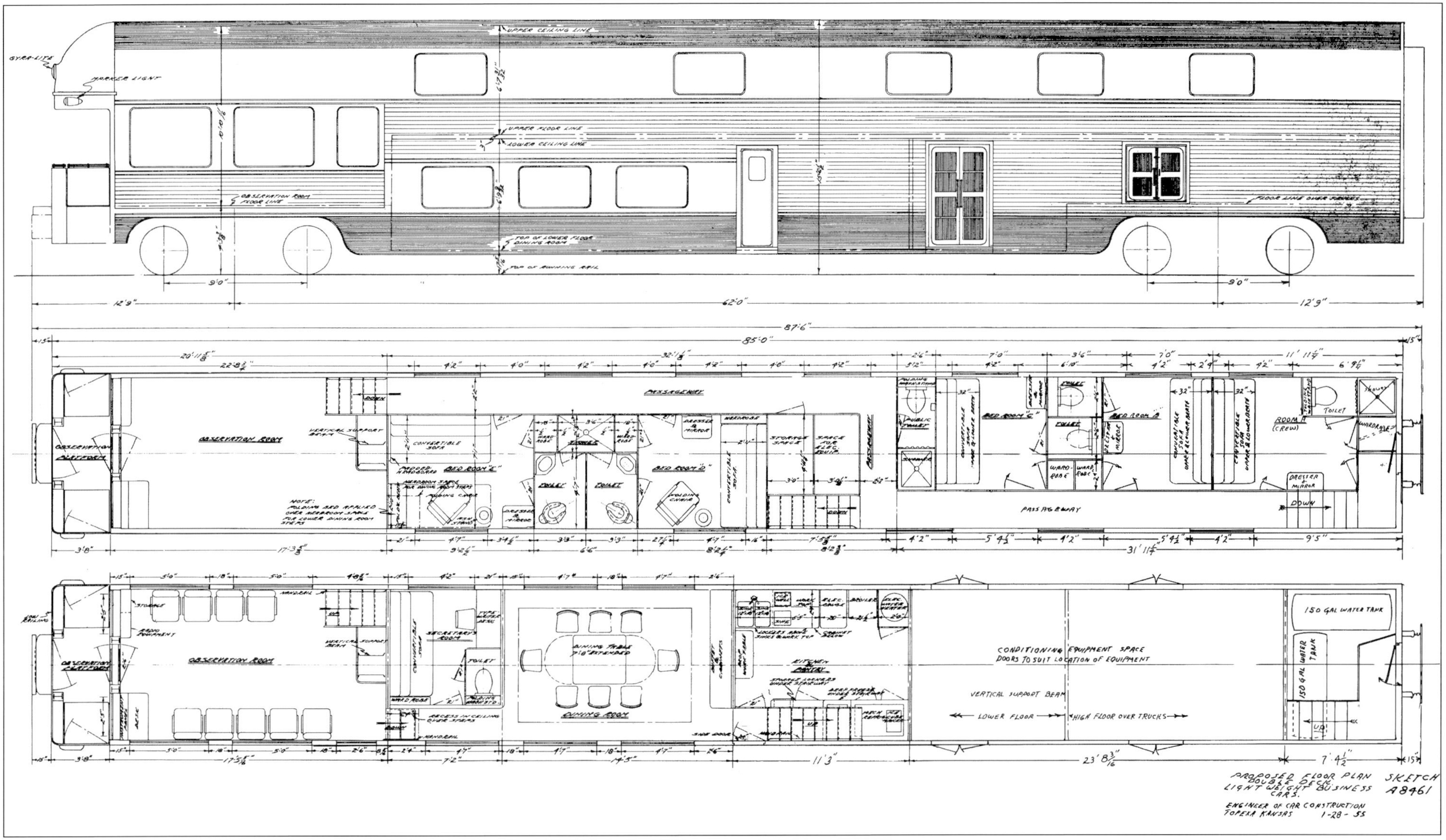

GYRA-LITE
MARKER LIGHT
UPPER CEILING LINE
UPPER FLOOR LINE
LOWER CEILING LINE
OBSERVATION ROOM FLOOR LINE
TOP OF LOWER FLOOR DINING ROOM
TOP OF RUNNING RAIL
FLOOR LINE OVER TRUCKS
9'0"
12'9"
62'0"
87'6"
85'0"
PASSAGEWAY
OBSERVATION ROOM
OBSERVATION PLATFORM
BED ROOM "E"
BED ROOM "D"
BED ROOM "C"
BED ROOM "B"
ROOM "A" (CREW)
CONVERTIBLE SOFA
TOILET
DRESSER & MIRROR
STORAGE SPACE
PASSAGEWAY
DOWN
SECRETARY'S ROOM
DINING TABLE 7'10" EXTENDED
DINING ROOM
KITCHEN & PANTRY
CONDITIONING EQUIPMENT SPACE
DOORS TO SUIT LOCATION OF EQUIPMENT
VERTICAL SUPPORT BEAM
LOWER FLOOR
HIGH FLOOR OVER TRUCKS
150 GAL WATER TANK
150 GAL WATER TANK
UP
11'3"
23' 8 3/16"
7' 4 1/2"
15"
PROPOSED FLOOR PLAN
DOUBLE DECK
LIGHT WEIGHT BUSINESS
CARS.
SKETCH
A8461
ENGINEER OF CAR CONSTRUCTION
TOPEKA KANSAS 1-28-55

Santa Fe purchased two identical ultra-modern 87-foot six-inch business cars from the Budd Company in 1957. They were named TOPEKA and 3rd SANTA FE. In the view above we see the SANTA FE ten years later at Dallas on October 14, 1967. The opposite side of the car is seen, right, at Topeka in May 1963. The Budd-built Santa Fe differed most notably from the Pullman built car of the same name with its six-wheel trucks and curved observation platform surround. Note the array of antennae on the roof and the Pyle Gyralite on the rear platform. *-Above, John B. McCall; right, F.O. Kelley, John B. McCall collection*

Detail views of the SANTA FE at Topeka, Kansas, in April 1976, showing placement of small "ATSF 50" in the letterboards at the front end of the car. A small vertical "SANTA FE," designating ownership, has also been placed beside the platform step handrail. The plate on the front door of the car reads "BUSINESS." At right is the small galley door. *-All Frank Ellington*

Both sides of nineteen year old Business car 3rd Santa Fe are seen in these two views at Topeka, Kansas, in April 1976. The car is still in essentially as-delivered condition. Three years later, on April 21, 1979, the interiors of the observation lounge in the car are still essentially as delivered. *-Above and right, Frank Ellington; interiors, Joe McMillan*

Car 3rd Santa Fe is on rear end of train No. 20, *The Chief,* at Pasadena, California, in early 1960. The Santa Fe was traditionally assigned to the President, and President Ernest S. Marsh can be seen here standing on platform nearest the middle as No. 20 departs for the East. *-Bob Melbo*

Twenty-seven year old Business car 3rd Santa Fe is spotted at Albuquerque, below, on February 11, 1984. Note the rear markers have been plated over. *-Bill Childers collection*

During 1990, Santa Fe re-equipped the Budd six-wheel cars, Topeka and Santa Fe, with four wheel trucks, presumably for a shorter wheel base, which would allow the cars to navigate industrial trackage. At the same time, the interiors of the cars were redecorated with wood paneling, muted colors and traditional Indian patterns to make the rather Spartan interiors reflect the heritage of the Santa Fe Railway. These two views, left, show the observation lounge in the Santa Fe at Chicago, December 28, 1990, after the work had been completed. *-Two photos, Joe McMillan*

The Santa Fe car is seen at San Diego, California, left, about 1992 after being fitted with four wheel trucks. Below is the opposite side view after the truck change in 1991. Originally assigned number 5003 for computer purposes, it was never applied to the car itself. Then, on September 30, 1973, it was renumbered 50, Santa Fe, at Chicago with the small ATSF 50 on the letterboard on both sides at the front of the car. After the BNSF merger, the car was retained and as of this writing is BNSF 7, Santa Fe. *-Left, Seeth Corky Thompson photo, SFRH&MS Collection; below, John B. McCall collection*

51 A.T.&S.F. TOPEKA

Brand new Business car TOPEKA arrives at Topeka, Kansas, its namesake city, above, on August 21, 1957, shortly after delivery from Budd. The TOPEKA, and sister SANTA FE were the last Business cars built new for the Santa Fe. *-F.O. Kelley photo, John B. McCall collection*

The two views at left show the TOPEKA in service some 15 years later at Tulsa, Oklahoma. *-Two photos, Railroad & Heritage Museum Collection, Temple, Texas*

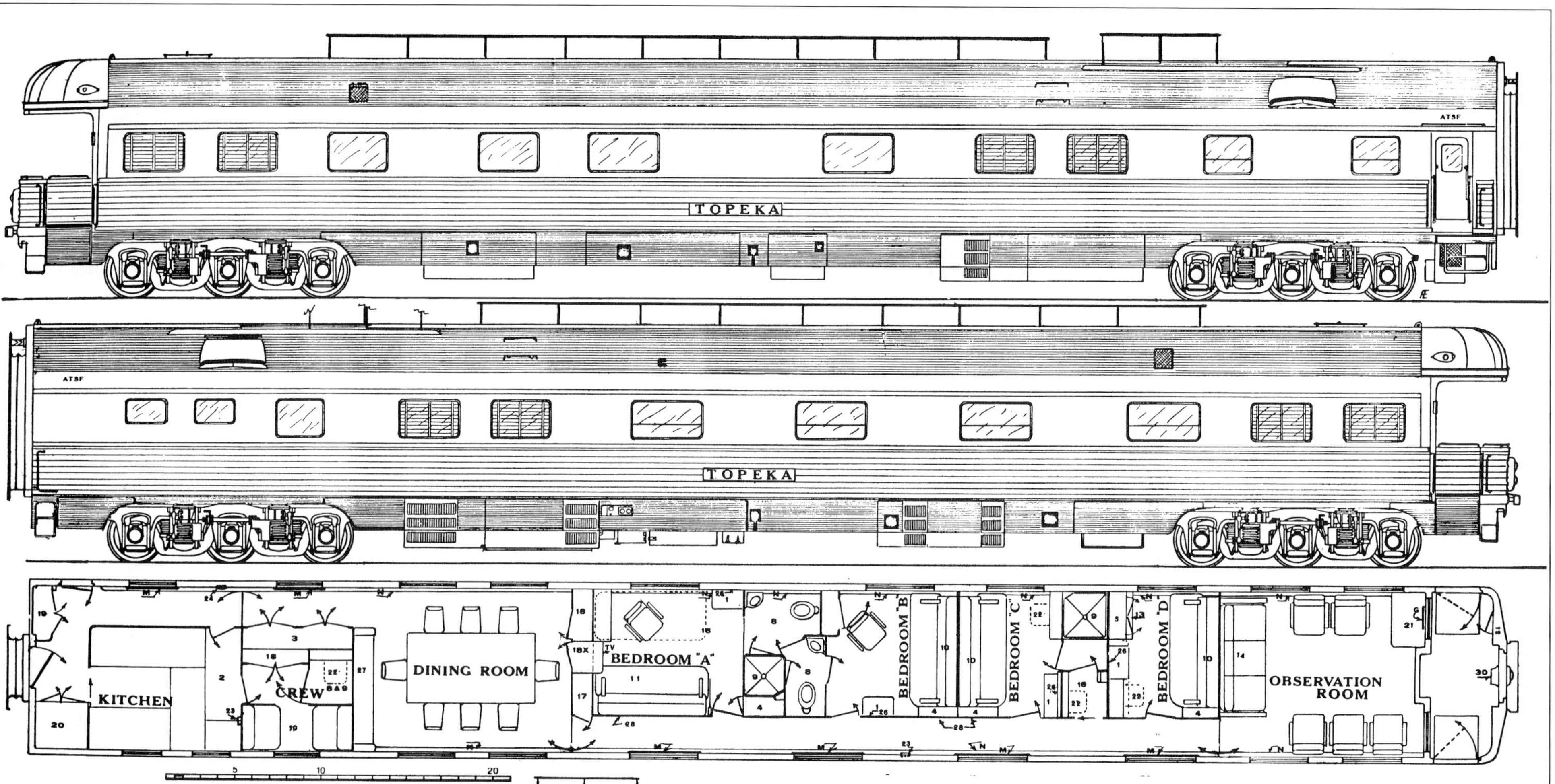

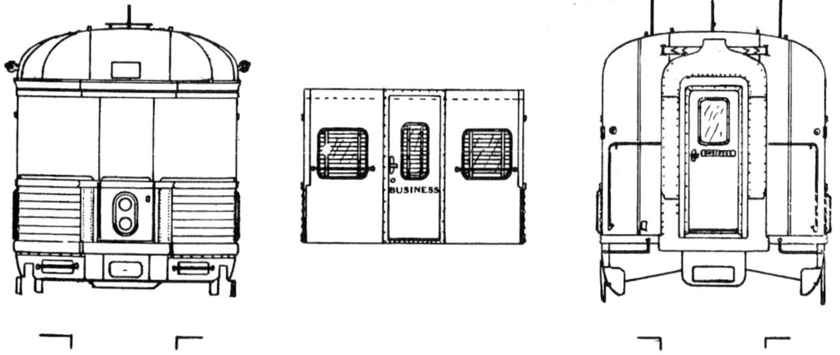

Business Car 51
Topeka

Drawn as measured Topeka, Kansas, June 16, 1973
Floorplan data drawing 23 C-4519 8/1/57
Drawn by Frank M. Ellington

KEY

1 Dresser
2 Pantry
3 Locker
4 Wardrobe
5 Overhead Cabinets
6 Heater Guard
7 Window Guard
8 Toilet
9 Shower
10 Upper and Lower Berth
11 Lower Berth Only
12 Backup equipment locker including conductor's and signal valves.
13 Secretary's Desk
14 Convertible Sofa
15 Folding Bed
16 General Toilet
17 Clean & soiled linen
18 Locker
18X TV-locker below
19 Switch locker
20 General Storage above, radio & music below
21 Desk w/ instrument panel
22 Folding Hopper
23 Chimes
24 Annunciator
25 Gyralite
26 Mirror
27 Buffet-Cabinets
28 Drawer below

The Topeka is seen at various points on the system: top Topeka at La Junta on the rear of the *Super Chief*, April 4, 1967. Private car Mt. Rubidoux keeps company with the Topeka at Grand Canyon, above, June 12, 1965. At left, the Topeka lays over at Phoenix in the winter of 1960. *-Top, Steve Patterson; above, Chard Walker photo, SFRH&MS Collection; left, Bob Knoll*

Car TOPEKA in service at San Diego on November 15, 1969. *-Seeth Corky Thompson photo, SFRH&MS Collection*

The interiors of the 1957 Budd-built cars, though sleek on the outside, lacked the rich interiors of early eras. The two views above show the dining room of the TOPEKA as delivered with rather plain inside with typical 1950s furnishings. Chicago, April 21, 1979. *-Two photos, Joe McMillan*

Left: The TOPEKA, designated car 51 on September 27, 1973, had by April 1979 large numbered plates affixed to sides of the car at the front. *-Frank Ellington*

Right: The Frederic Mizen painting "Indian Chant," which appeared on the 1953 Santa Fe calendar, adds a touch of class in the observation lounge, but the fabrics, wall paneling and colors in the TOPEKA were spartan at best. *-Joe McMillan*

Just fitted with four-wheel trucks, the Topeka is shown, above, in June 1990. The two interior views at Chicago, December 28, 1990, show the redecorating done in the observation lounge and dining room. After the BNSF merger, the car was retained and as of this writing is BNSF 6, Topeka. *-Above, Louis A. Marre, John B. McCall collection; others, Joe McMillan*

Atchison

Pullman built the first stainless steel Business car on the system, the 2nd Santa Fe, in 1949. It was renamed Atchison when Budd delivered the 3rd Santa Fe in 1957. The Pullman-built Santa Fe is seen, left, on rear of No. 123, the *Grand Canyon*, at San Bernardino depot in October 1952. That's Bill Whitmeyer, son of photographer Jack Whitmeyer, standing alongside the attendant. The Santa Fe is seen a few years later, below, at Santa Fe Springs. Note that a Pyle Gyralite has replaced the single light on the platform gate, and different style markers have been applied. *-Left, Jack Whitmeyer, SFRH&MS Collection; below, Donald Duke*

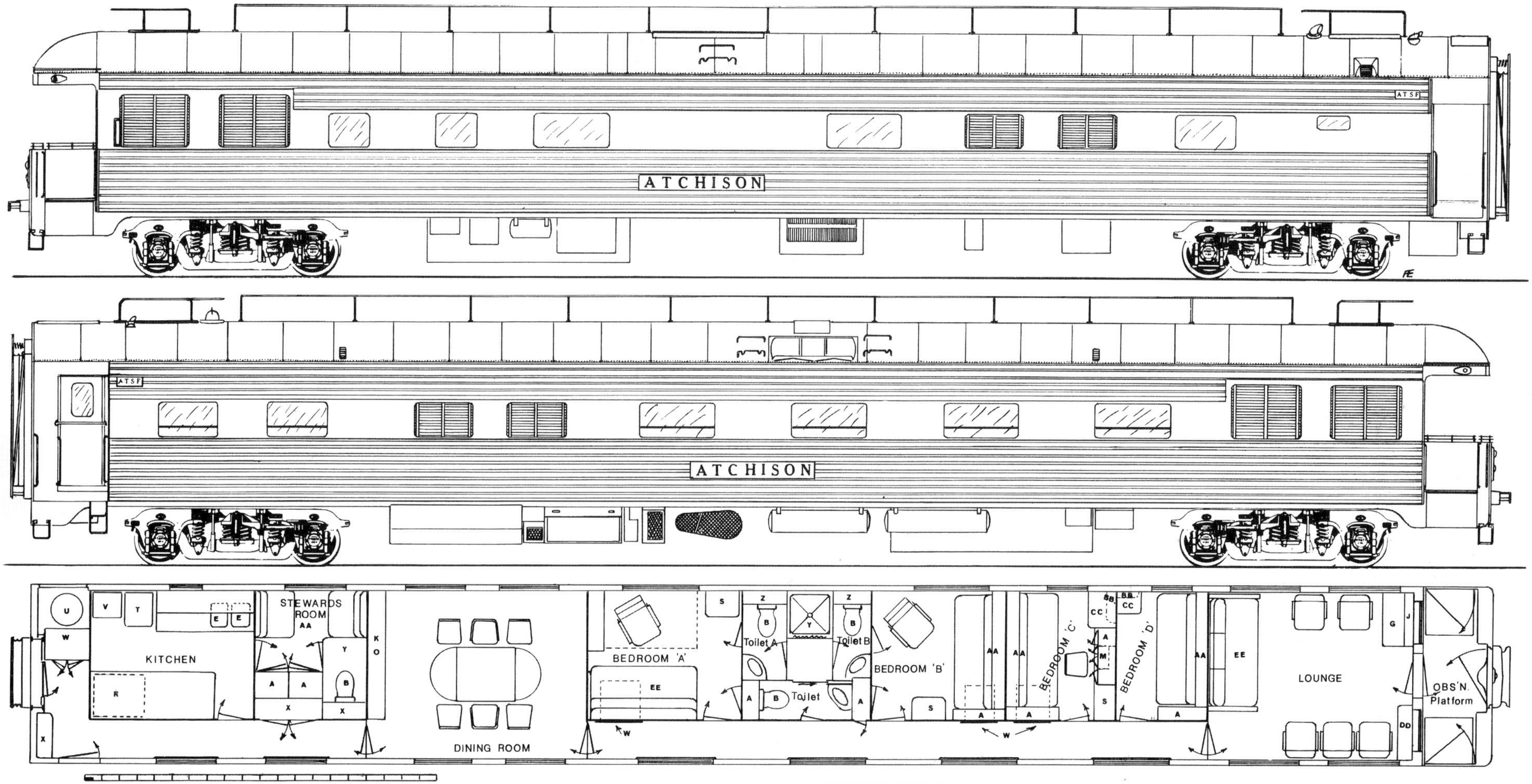

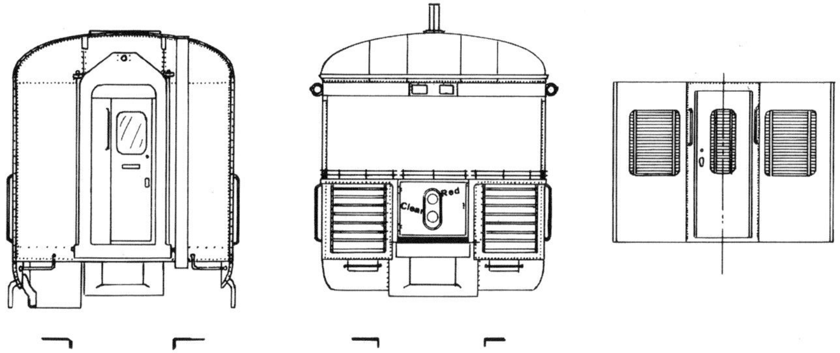

Business Car 52 Atchison

Drawn as measured Chicago, December 23, 1969
Floorplan data drawing 23 C-3747 6/19/59
Drawb by Frank M. Ellington

KEY

A	Wardrobe	I	Chime	R	Refrigerator	Z	Shelves Overhead
B	Hopper	J	Speedometer	S	Dresser	AA	With Upper Berth
C	Washstand	K	Sideboard	T	Range	BB	Folding Washstand
D	Ice Box	L	Annunciator	U	Water Heater	CC	Folding Hopper
E	Sink	M	Secretary's Desk	V	Charcoal Broiler	DD	Radio
F	Drainboard	N	Bookcase	W	Linen/Storage	EE	Convertible Sofa
G	Table	O	Buffet	X	Electric Reg. Locker		
H	Switchboard	P	Locker	Y	Shower		

The ATCHISON prepares to depart Topeka, Kansas, top, September 7, 1968, and is seen, right, spotted at Slaton, Texas, January 20, 1964. By April 1976, the ATCHISON, now nearly 30 years old, is seen below at Topeka, Kansas, sporting a small "SANTA FE" plate on the platform gate above the Gyralite. *-All, Joe McMillan*

Interior views of the ATCHISON taken at Chicago, April 21, 1979, reflect the rather plain interiors that characterized Santa Fe's lightweight Business cars as delivered. Upper left is the dining room looking toward the kitchen and crew quarters. Upper right we see the observation lounge with desk and instrument panel. At left is the stainless steel-paneled kitchen looking toward the front of the car. At right is Bedroom "B" as viewed from the window toward the passageway door. Both the sofa and the folding upper berth are made up in this view. Note the smooth, light colored dresser. *-Color views, Joe McMillan; others Santa Fe Railway, Frank Ellington collection*

The ATCHISON is seen trailing a freight train at Edison, California, in February 1990. Note absence of marker lights and small identification plate near the galley door. *-Louis Marre, John B. McCall collection*

The ATCHISON, along with the other lightweight cars in the fleet, was redecorated in 1990 with wood paneling, dark wood cabinets, more contemporary colors and soffit lighting. These views of the car were taken at Chicago December 28, 1990. From left we see the observation lounge convertible sofa, the dining room, and the observation lounge facing the rear of the car with small desk and instrument cluster. *-Three photos, Joe McMillan*

53 A.T.&S.F. Mountainair

Santa Fe purchased its only secondhand lightweight Business car from United States Steel Corp. in August 1962. The car, while owned by U.S. Steel was named Laurel Ridge, and it emerged from Topeka Shops renamed AT&SF Mountainair in November 1962. Here we see the Mountainair in original exterior configuration on No. 1 at Clovis, New Mexico, February 24, 1963. The rounded roof, large red oscillating light and lack of observation platform made this car unique. This is also believed to be the only car on the system, which had stainless steel ribbing along with shadow striping. *-John B. McCall*

Six months after arriving on the property, the Mountainair brings up the rear of No. 20, *The Chief,* at Summit on May 26, 1963. By April 28, 1966, the rear of the Mountainair, seen at Topeka, right, had been modified with a conventional diaphragm, and the rounded roof and giant red light were gone. *-Above, Chard Walker, SFRH&MS Collection; right Steve Patterson*

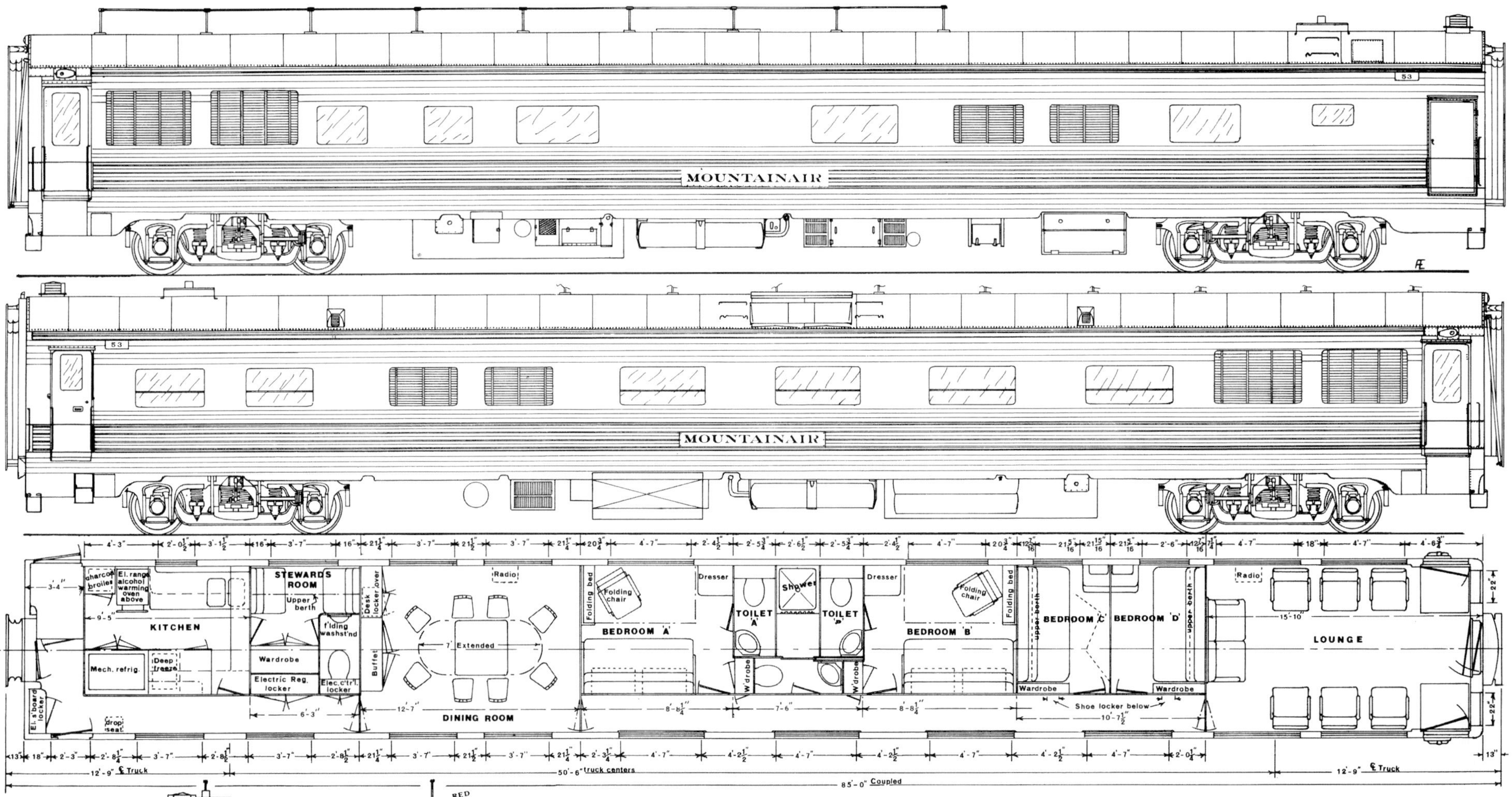

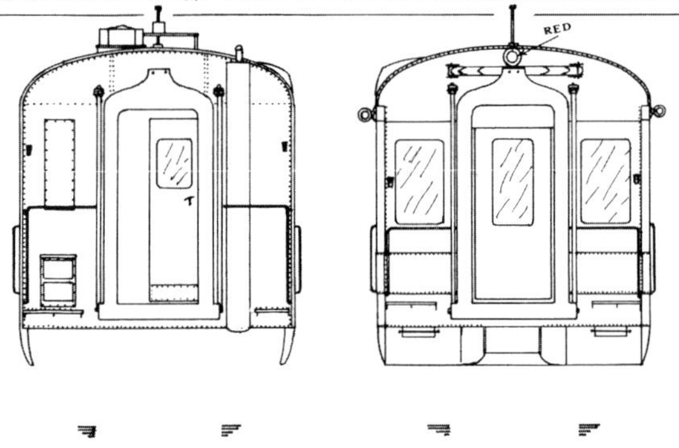

Business Car 53
Mountainair

Floorplan data drawing 23 C-4792 12/4/62
Redrawn by Frank M. Ellington

KEY

1 Dresser
2 Pantry
3 Locker
4 Wardrobe
5 Overhead Cabinets
6 Heater Guard
7 Window Guard
8 Toilet
9 Shower
10 Upper and Lower Berth
11 Lower Berth Only
12 Backup equipment locker including conductor's and signal valves.
13 Secretary's Desk
14 Convertible Sofa
15 Folding Bed
16 General Toilet
17 Clean & soiled linen
18 Locker
18X TV-locker below
19 Switch locker
20 General Storage above, radio & music below
21 Desk w/ instrument panel
22 Folding Hopper
23 Chimes
24 Annunciator
25 Gyralite
26 Mirror
27 Buffet-Cabinets
28 Drawer below

Detail views of ATSF 53, Mountainair, at Chicago 18th St. coachyard on April 7, 1973. The two views at left are both sides of the front. Note the small ATSF initials, the only identifying markings showing ownership. *-All, Frank Ellington collection*

Both sides of Car 53, the MOUNTAINAIR, are seen at Chicago, above, October 10, 1983, and Dallas, right, in August 1985. By this time, the car had lost its shadow striping and was completely self contained with its own power generating equipment. Small plates at both ends identify the car as ATSF 53. Note that the trucks have been changed. *-Above, Joe McMillan; right, Bill Childers*

The rather dated interiors of the Mountainair, as exemplified by the dining room, above left, at Chicago, April 21, 1979, were completely redone in 1990 with dark wood paneling and cabinets, and fabric designs and colors with a southwest motif. *-All, Joe McMillan*

Santa Fe's Director's train rolls through the Kansas countryside in 1991 behind a pair of Warbonnet GP60s. -*Courtesy Kevin Eudaly*

7
Director's Cars

When Amtrak assumed operation of Santa Fe's passenger service in May 1971, most of Santa Fe's passenger rolling stock was so modern and well maintained that Amtrak actively sought after and purchased large blocks of this equipment. During the transactions, Santa Fe wisely retained a few cars for future use. In addition to the Business car fleet, these included a few Sleepers, a Dome, a Lounge, a Diner, some Baggage cars and the like. Whereas previously when Santa Fe needed this equipment in conjunction with an official move, they could apply to their own coach yards for the necessary cars. This would no longer be the case. These cars retained by Santa Fe were stored at San Bernardino and Topeka.

At last, during the last few months of 1974 and the first few months of 1975 it became obvious that a very special train would be operated that year for the board of directors to inspect certain portions of the system prior to their annual meeting, a practice that the railroad had used frequently in the 1950s.

First, Lounge-Diner 1397 was stripped and entered the Topeka passenger car shops to begin its conversion to one of the system's most unique inspection cars.

The six REGAL series (4 compartment, 4 double bedroom, 2 drawing room) sleeping cars were taken out of storage at San Bernardino, to be serviced and refurbished.

During early April things began to come together. Steam Generator car 139, Chair-Lounge-Dome 60, Diner 61, and Club-Lounge 62 were added to the line of Sleeping cars at San Bernardino. The ex-Lounge-Diner, now numbered 89, was nearing completion as a Track Inspection car at Topeka.

Tuesday evening, April 1, 1975, Cars 60 through 68 and Steam Generator car 139 were sent on a shake-down run from San Bernardino to Los Angeles on the 198 train, the westbound *Super C*, and returned to San Bernardino the following morning on the 891 train, the eastbound *Super C*.

The Steam Generator car and passenger cars were dropped at San Bernardino for inspection and any minor repairs. It was at this time that small car numbers were stenciled near the vestibule door on the exterior of the Sleeping cars, 63 through 68.

Wednesday morning, April 9, 1975, the 891 train from Los Angeles, picked up the *Directors Special* cars at San Bernardino and forwarded them to Chicago in a deadhead move. The consist between San Bernardino and Chicago was as follows:

Four SD45-2s
139 Steam Generator
61 Diner
67 REGAL MANOR
68 REGAL SPA
64 REGAL HUNT
65 REGAL LANE
63 REGAL CREST
66 REGAL LARK
60 Chair-Lounge-Dome
12 TOFC flat cars
Caboose (Ce-1)

Meanwhile, mechanical department forces in the passenger car shops at Topeka had completed Theater-Lounge-Track Inspection car 89 out of former Lounge-Diner 1397, and it was taken to Chicago as well.

Without doubt, the most unique of the Directors cars was ATSF 89, the Theater-Lounge-Track Inspection car, the rear of which is seen in service at Kansas City in May 1988. *-Mike Martin*

Everything was brought together in Chicago in preparation for the first use of the Directors Special. Even FP45 5942 was brought out of storage at Barstow and forwarded to Chicago on April 14th. It was felt that the aesthetic lines of the FP45, with the new yellow "bonnet" paint scheme, would blend well with the new special train.

At last, on the afternoon of April 27, 1975, the new equipment went into service for the first time as the Directors Special departed Chicago for the first leg of its journey with a swing through Texas. The consist was:

5942 (FP45), 5708, 5709 (SD45-2s)
139 Steam Generator
52 Atchison
53 Mountainair
66 Regal Lark
67 Regal Manor
63 Regal Crest
68 Regal Spa
64 Regal Hunt
60 Chair-Lounge/ Dome
61 Diner
51 Topeka
50 Santa Fe
89 Track Inspection car

As can be seen by the above consist, the four lightweight Business cars, Atchison, Topeka, Santa Fe and Mountainair formed an integral part of the Director Cars set and were used as needed. Later in 1976, stored Baggage cars 3923, 3906 and 3914 were reconditioned and renumbered ATSF 74, 75 and 77, respectively, for use with the Directors cars and other official business.

For the next 20 years, these cars were used together in their entirety, or in smaller groupings to support other special movements, such as the Annual Tournament of Roses parade trains, shipper's specials etc.

At the time of the BNSF merger, all of these cars, except ATSF 63, Regal Crest, were retained for business use and are in service as of this writing. Two cars were renamed, honoring their Santa Fe heritage: Dining car 61, renumbered BNSF 11, was named Fred Harvey, and Theater-Lounge-Track Inspection car 89, renumbered BNSF 32, was named William B. Strong.

SANTA FE DIRECTORS CARS

Car Name and Number	*Type*	*Builder Date Built*	*Length Trucks*	*Notes & Disposition*
ATSF 60	Chair-Lounge-Dome	Budd 1954	85' 6wh	Originally ATSF Big Dome 506, renumbered ATSF 60, 1975; to BNSF 31, Bay View, 1998.
ATSF 61	Diner	Pullman 1950	85' 4wh	Originally Diner ATSF 600, renumbered ATSF 61, 1975; named Fred Harvey prior to 1990; to BNSF 11,1998, and name Fred Harvey retained.
ATSF 62	Club-Lounge	Pullman 1947	85' 4wh	Originally ATSF 1348, Club-Lounge; renumbered ATSF 62,1975; to BNSF 29, Valley View, 1998.
ATSF 63 Regal Crest	Sleeper	AC&F 1950	85' 4wh	Originally 4-4-2 sleeper Regal Crest; assigned number ATSF 63, 1975; to BNSF 81, Track Measurement Support, 1998.
ATSF 64 Regal Hunt	Sleeper	AC&F 1950	85' 4wh	Originally 4-4-2 sleeper Regal Hunt; assigned number ATSF 64, 1975; to BNSF 64, Marias Pass, 1998.
ATSF 65 Regal Lane	Sleeper	AC&F 1950	85' 4wh	Originally 4-4-2 sleeper Regal Lane; assigned number ATSF 65, 1975; to BNSF 65, Raton Pass, 1998.
ATSF 66 Regal Lark	Sleeper	AC&F 1950	85' 4wh	Originally 4-4-2 sleeper Regal Lark; assigned number ATSF 66, 1975; to BNSF 66, Cajon Pass, 1998.
ATSF 67 Regal Manor	Sleeper	AC&F 1950	85' 4wh	Originally 4-4-2 sleeper Regal Manor; assigned number ATSF 67, 1975; to BNSF 67, Donner Pass, 1998, renamed Trinchera Pass, 2003.
ATSF 68 Regal Spa	Sleeper	AC&F 1950	85' 4wh	Originally 4-4-2 sleeper Regal Spa; assigned number ATSF 68, 1975; to BNSF 68, Rollins Pass, 1998.
ATSF 74	Baggage	Topeka 1965	73'11" 4wh	Originally Bagg. & Exp. ATSF 3923, built Topeka from Pullman shell, 1963; renumbered ATSF 74, 1976.
ATSF 75	Baggage	Topeka 1963	73'11" 4wh	Originally Bagg. & Exp. ATSF 3906, built Topeka from Pullman shell, 1963; renumbered ATSF 75, 1976.
ATSF 77	Baggage	Topeka 1965	73'11" 4wh	Originally Bagg. & Exp. ATSF 3914, built Topeka from Pullman Shell, 1965; renumbered ATSF 77, 1976; to BNSF 77, 1998.
ATSF 89	Theater-Inspection	Budd 1940	79'8" 4wh	Originally ATSF 1397, Club-Lounge, rebuilt to Diner-Lounge 8/52; rebuilt as Theater-Inspection car and renumbered ATSF 89, 1975; to BNSF 32, William B. Strong.

Frequently part of the consist of *The Chief* and other Santa Fe streamliners, Chair-Lounge-Dome ATSF 506, built by Budd in 1954, was retained by the Santa Fe in 1971 and was used quite extensively prior to the Directors Special in 1975. At left, the car is seen at Topeka in July 1975 several months after it had been integrated into the Directors Car fleet as ATSF 60. In the view above, taken at Topeka in August 1988, we see the opposite side of the car. The steel bars at each end of the dome were to prevent icicles at tunnel portals or bridges from breaking dome windows and causing injury to passengers. It was retained by the BNSF after the merger, renumbered to BNSF 31 in 1998, and named Bay View. As of this writing the car is still in service.
-Two photos, Frank Ellington

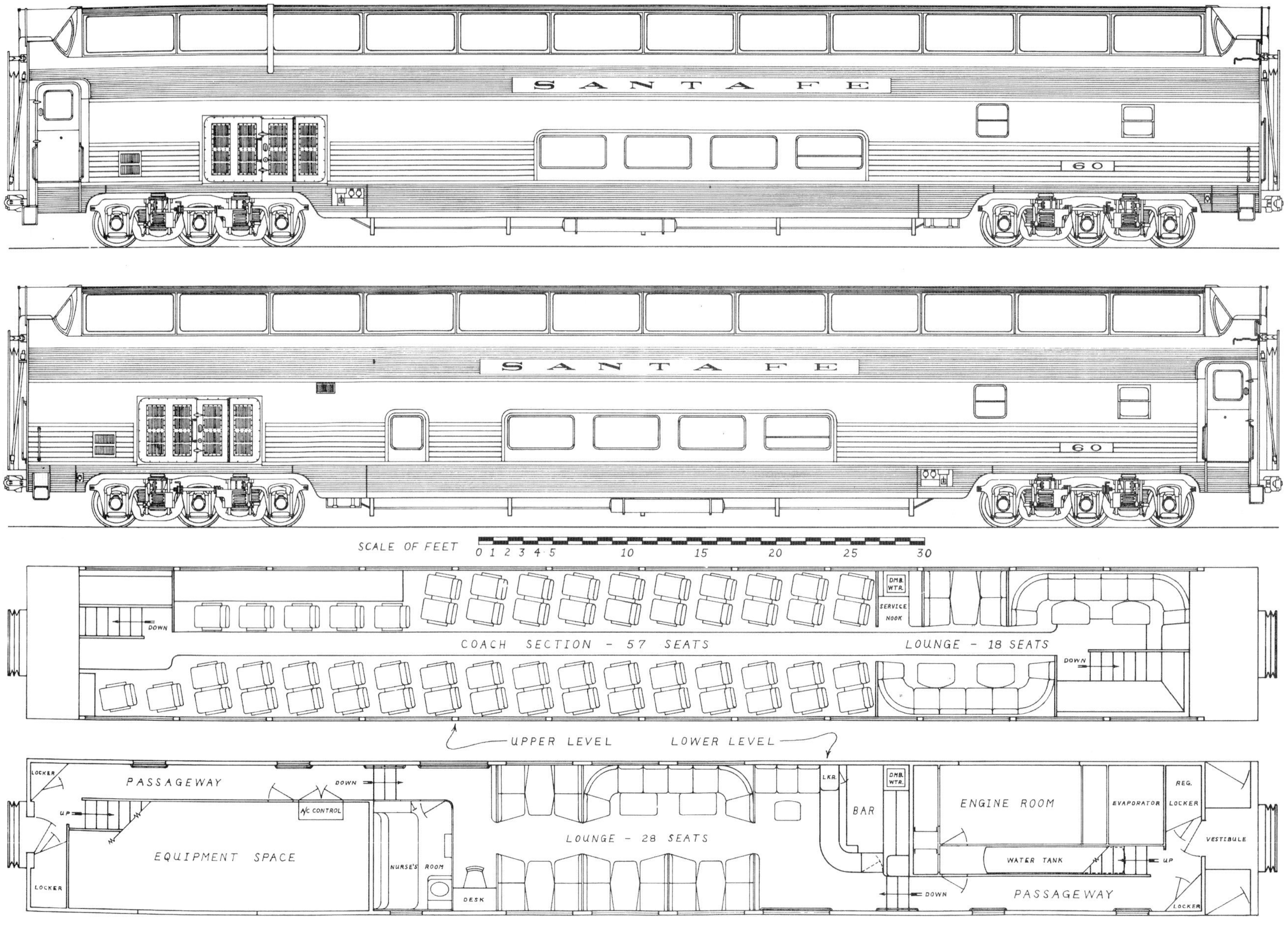

Dome-Lounge Car 60

Budd 1954

Former ATSF 506, circa 1973

Drawn by Dennis J. Kogan

61 A.T.&S.F. Fred Harvey

At the time Amtrak assumed operation of the nation's railroads in May 1971, Santa Fe wisely retained 85-foot Diner ATSF 600, built by Pullman in 1950 for the *Super Chief*, for business use. As the Directors Car set came together in late 1974-early 1975, the 600 was renumbered ATSF 61. At a later date, as early as 1990, the car was also named Fred Harvey. Here we see both sides of the car at Topeka in June 1990. The lower view is the kitchen side. The 61 was retained by the BNSF after the merger and renumbered to BNSF 11 in 1998. BNSF retained the name Fred Harvey, and the car remains in service as of this writing. *-Right, Lee Burglund photo, SFRH&MS Collection; below, Bill Childers*

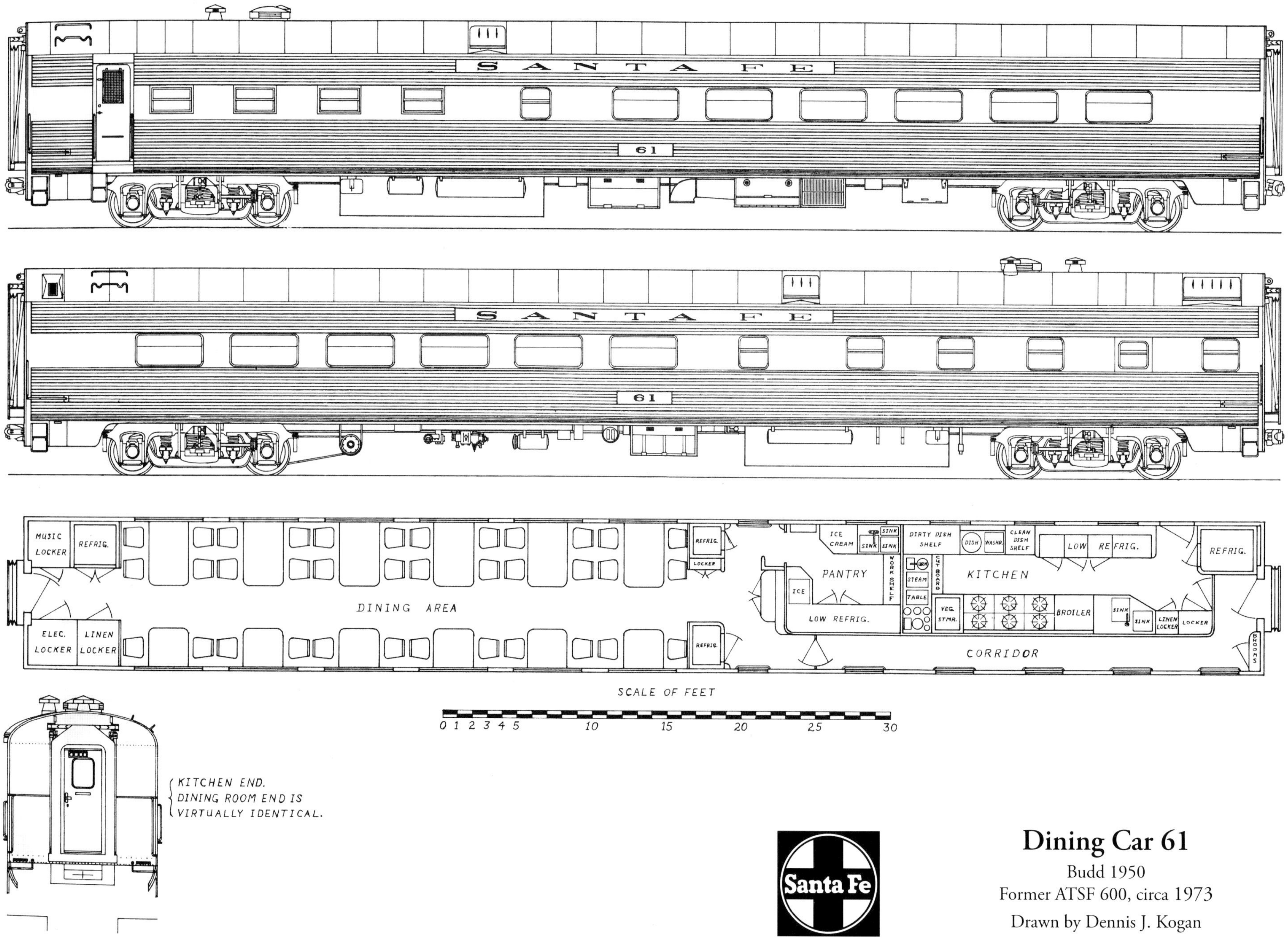

Dining Car 61

Budd 1950

Former ATSF 600, circa 1973

Drawn by Dennis J. Kogan

62
A.T.&S.F.

Pullman 1947-built 85-foot Club-Lounge ATSF 1348 was also retained by the Santa Fe for business use and was renumbered ATSF 62 in 1975. Unlike Big Dome 60, which had self-contained generators for its air conditioning and heating, cars 61 and 62 had to operate with a Steam Generator car when in use. Accordingly, Club-Lounge 62 is seen with a Steam Generator car on the rear of the *Super C*, right, at Gale, California, on the Needles District of the Los Angeles Division, on January 13, 1978. By August 1988, ATSF 62, seen below at Topeka, had been re-trucked and fitted with Amtrak headend power connections. Note unusual red markers on the rear bulkhead walls. ATSF 62 was retained after the merger, was renumbered BNSF 29, and named Valley View in 1998. The car is still in service as of this writing.
-Right, Phil Kauke; below, Frank Ellington

Regal Crest

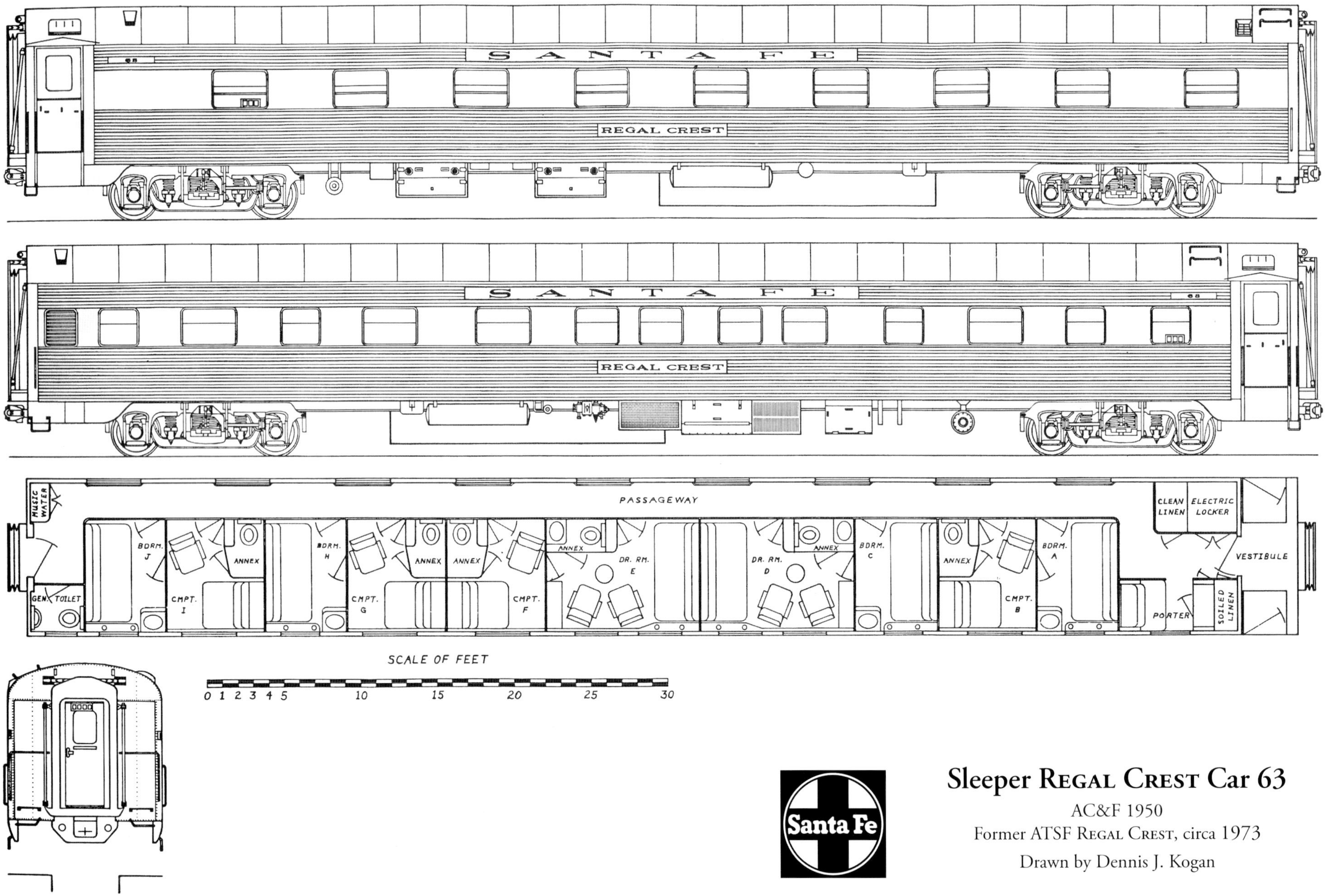

Sleeper Regal Crest Car 63

AC&F 1950

Former ATSF Regal Crest, circa 1973

Drawn by Dennis J. Kogan

64 A.T.&S.F. Regal Hunt

ATSF Sleeper 64, Regal Hunt, at Chicago, May 21, 1983, was built as an 85-foot 4-4-2 sleeper Regal Hunt by AC&F in 1950. It was retained by the Santa Fe after Amtrak and assigned number ATSF 64 in 1975. After the merger it was retained by the BNSF, became BNSF 64, and named Marias Pass in 1998. It is still in service as of this writing. *-Joe McMillan*

65 A.T.&S.F. Regal Lane

ATSF Sleeper 65, Regal Lane, at Kansas City, February 18, 1990, was built as an 85-foot 4-4-2 sleeper Regal Lane by AC&F in 1950. It was retained by the Santa Fe after Amtrak and assigned number ATSF 65 in 1975. After the merger it was retained by the BNSF, became BNSF 65, and named Raton Pass in 1998. It is still in service as of this writing. *-Joe McMillan collection*

66 A.T.&S.F. Regal Lark

ATSF Sleeper 66, Regal Lark, at Chicago, May 21, 1983, was built as an 85-foot 4-4-2 sleeper Regal Lark by AC&F in 1950. It was retained by the Santa Fe after Amtrak and assigned number ATSF 66 in 1975. After the merger it was retained by the BNSF, became BNSF 66, and named Cajon Pass in 1998. It is still in service as of this writing. *-Joe McMillan*

67 A.T.&S.F. Regal Manor

ATSF Sleeper 67, Regal Manor, at Chicago, April 20, 1985, was built as an 85-foot 4-4-2 sleeper Regal Manor by AC&F in 1950. It was retained by the Santa Fe after Amtrak and assigned number ATSF 67 in 1975. After the merger it was retained by the BNSF, became BNSF 67, and named Donner Pass in 1998. It was renamed Trinchera Pass in 2003. It is still in service as of this writing. *-Joe McMillan*

68 A.T.&S.F. Regal Spa

ATSF Sleeper 68, Regal Spa, at Chicago, May 21, 1983, was built as an 85-foot 4-4-2 sleeper Regal Spa by AC&F in 1950. It was retained by the Santa Fe after Amtrak and assigned number ATSF 68 in 1975. After the merger it was retained by the BNSF, became BNSF 68, and named Rollins Pass in 1998. It is still in service as of this writing. *-Joe McMillan*

77 A.T.&S.F.

Baggage car ATSF 77 was originally 73-foot 11-inch Baggage & Express car ATSF 3914, built at Topeka Shops in 1965 from a shell supplied by Pullman. It was renumbered ATSF 77 in 1976, and is seen here at Cadiz, California, on a Rail-Cycle Special in March 1991. It was retained by the BNSF after the merger and was became BNSF 77 in 1998. *-Mike Martin*

ATSF Theater Inspection car 89 awaits the call to service at Topeka in May 1979. Originally Club-Lounge then Lounge-Diner ATSF 1397, built by Budd in 1940 for the *Super Chief,* the 79-foot, 8-inch car was retained by the Santa Fe after Amtrak began operations and stored at Topeka. Over the winter of 1974-1975, mechanical department forces in the passenger car shops at Topeka stripped the car to begin its conversion to one of the system's most unique inspection cars, emerging in April 1975 as Theater-Track Inspection car ATSF 89. Car 89 retained its fluted stainless steel exterior and is seen here in its earliest days, before being retrofitted with Budd outside swing hanger trucks. After the merger, it was renumbered to BNSF 32 and named William B. Strong in honor of the legendary system president from the Santa Fe's earliest days. As of this writing the car remains in service. *-Louis Marre photo, John B. McCall collection*

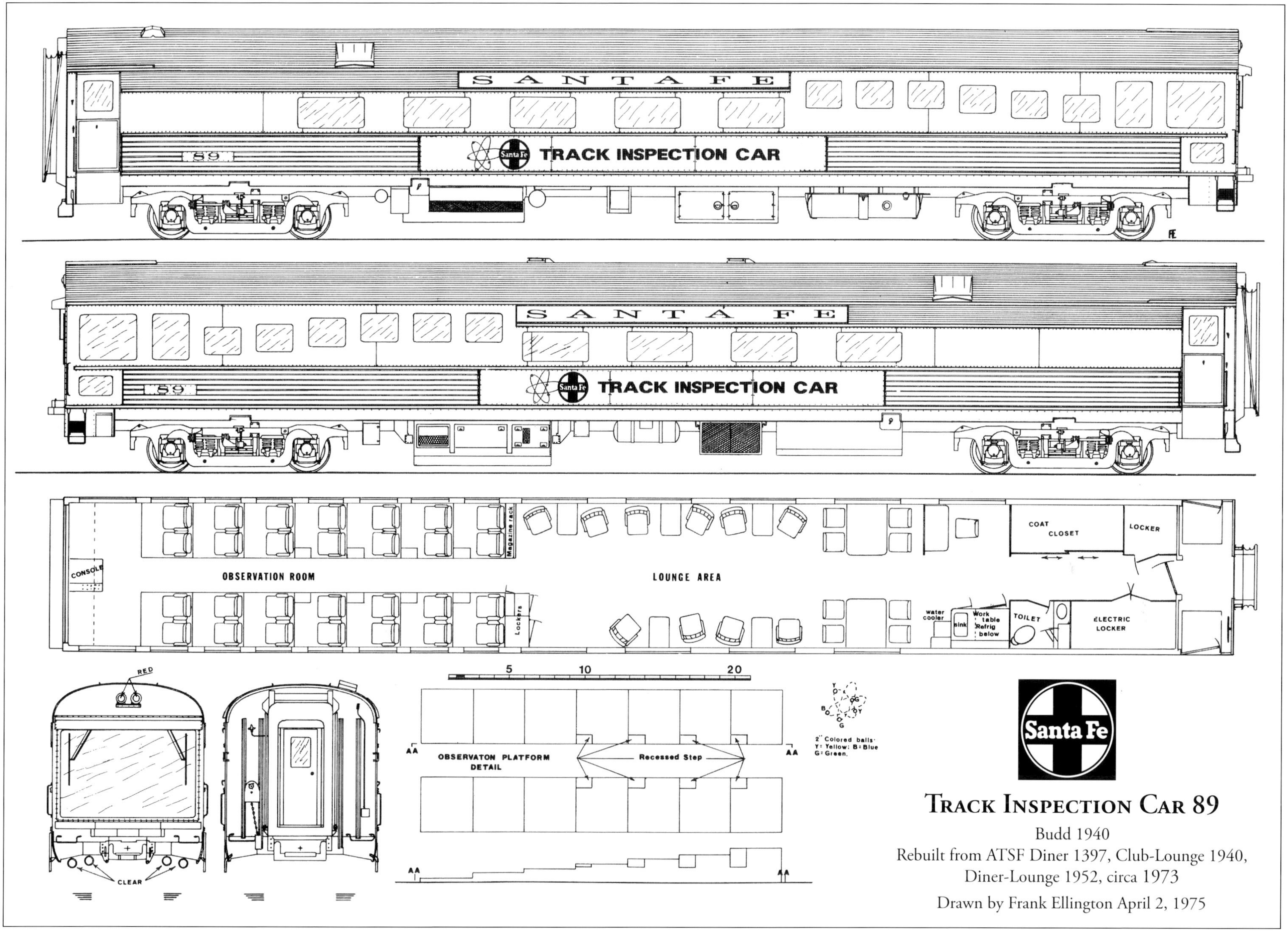

Track Inspection Car 89

Budd 1940

Rebuilt from ATSF Diner 1397, Club-Lounge 1940, Diner-Lounge 1952, circa 1973

Drawn by Frank Ellington April 2, 1975

These interior photos were taken at Chicago in April 1975 as car 89 made its debut prior to the operation of the Directors Special. At upper left we see the lounge area looking toward the front of the car. The light colored "runner" showing on the aisle in these photos is a drop cloth which is removed when the car is in service. In the view above we are looking at the lounge area, as viewed from the front of the car, with the theater area at the far end. At left we are in the observation section looking toward the rear showing the theatre-type seating. A protective metal screen is in place temporarily over the wide expanse of glass on the rear and is stored elsewhere when the car is in operation. This screen was later replaced by solid metal sheeting. As built, the car had a capacity of 48 people; 28 in the observation area and 20 in the lounge. -*Three photos, Joe McMillan*

View from the rear theater of car 89 in service. -*Railroad & Heritage Museum Collection, Temple, Texas*

Multi-panel protective metal sheets in place on car 89 to protect the theater glass window in storage and in deadhead movements. San Diego, May 1984. *-Mike Martin*

Originally heavy metal screens were used to protect the theater glass window, as seen here at Topeka in April 1975. The four panels were slid into place in slots and secured by pad locks. *-Mike Martin*

Car 89 in service on the Rose Parade Special at Pasadena, California, in January 1985. Note the outside swing hanger trucks. *-Mike Martin*

Santa Fe
SANTA FE RAILWAY

8

Special Purpose Cars

For a railway system as large and as old as the Santa Fe, it found itself in need of educating and training its employees as technology advanced. More over, it needed ways to monitor motive power, track infrastructure and other systems in order to make sound business decisions. This training and analysis needed to be mobile in order to bring it to the far reaches of the system, which resulted in a wide variety of interesting special purpose cars.

As early as 1900, Air Brake Instruction cars brought the latest in air brake development directly to division terminals. Similarly, steam power plant training and efficiency was mobile on the Santa Fe, as well as Rules Examiners and the need for regular Physical Examination of employees. As technology advanced, Santa Fe introduced a Diesel Locomotive Instruction car and a Radio Test car. The apex of the mobile training car was Locomotive Simulator, car 5008 (70). Introduced in September 1969, the $750,000 Santa Fe simulator was the first such training device to appear on any U.S. railroad, and was housed in one of a three-car training unit that could travel to the far reaches of the Santa Fe system via its own special train. As part of this unit, Santa Fe built up a Film car for taking movies of real segments of the Santa Fe which could be used in the simulator training environment.

As early as 1910, Santa Fe had its own Dynamometer car, No. 29, for testing the proficiency of their locomotives. Similarly, Track Recording cars evaluated the track. This led to the sophisticated Track Geometry car of later years.

In later years, as public relations became a valuable tool, Santa Fe introduced a number of Display and Exhibit cars. The railway even had a Motion Picture Film car, made out an old flat, stationed in Southern California.

Another area of special purpose cars was the Steam Generator cars, first converted from obsolete tenders taken from 1337 Class Pacifics during World War II, then in the 1970s from lightweight Baggage cars. By 1985, Santa Fe had discarded all but two of their Steam Generator cars, which were kept on hand to provide steam for the company business train.

Unfortunately we can only present a fragmented and incomplete roster of Santa Fe's special Purpose cars, as records are difficult to obtain. Virtually all of these cars were converted from other, obsolete types of cars. Most of these cars were numbered into the 5000 series beginning about 1925. During 1973, surviving cars from old series 5000s were renumbered into a scheme starting with number 70.

A few of these cars survive as of this writing on the roster of the BNSF. Track Geometry car 85 is as of this writing BNSF 85, Track Measurement car, and its support car ATSF 86 is currently BNSF 86, Track Measurement Support. Former Santa Fe Dynamometer 83 is currently BNSF 83, Research & Test car, named Canyon Diablo.

Opposite: A model poses with Display car 87, after it was redecorated in September 1973 at Topeka. This was originally Smoker ATSF 2952 built by Pullman in May 1930. Rebuilt several times, it was last Lounge-Dorm ATSF 1527, prior to being converted to Display car 2nd 5005 at Topeka in June 1967. *-Santa Fe Collection, Kansas State Historical Society*

Without doubt, the most grotesque face of Santa Fe's Special Purpose cars was that of Film Car 5009, converted from the heavyweight Business car 30. Renumbered Car 73, in 1973, it is seen here at Topeka, August 5, 1985, in its final livery of gray and silver. *-Dean Hale photo, Bill Childers collection*

SANTA FE SPECIAL PURPOSE CARS

Car No.	*Type*	*Builder Date Built*	*Length/ Trucks*	*Notes & Disposition*
20	Track Recording car	B&S 1886	60' 6wh	Converted from Business car ATSF 20, 1/27.
21	Radio Test car	B&S 1890	65' 6wh	Converted from Business car ATSF 21, 1940s; sold Stockton Terminal & Eastern, 1958.
24	Steam Heat Instruction	B&S 1900	70' 6wh	Converted from Diner ATSF 1412, 10/17; to Coach Smoker AT&SF 789, 1/26.
26	Air Brake Instruction		50' 4wh	Converted from 900 series Postal, 1903; renumbered 199905, 1912; renumbered 26, 1918; out by 1923.
2nd 26	Air Brake Instruction	Pullman 1907	75' 6wh	Converted from Parlor Observation ATSF 1203, 11/23; out by 1930.
27	Air Brake Instruction	Pullman 1887	51' 4wh	Converted from 2500 series Coach, 1900; renumbered 199906, 1912; renumbered 27, 1918; to 199169, Bunk car, 7/28/26; dismantled, 12/15/36.
2nd 27	Air Brake Instruction	Pullman 1912	70' 6wh	Converted from Smoking car ATSF 567, 6/26; retired, 7/61; sold to Ehrman Howell, 8/10/61.
28	Dynamometer Equipment		50' 4wh	Converted from Baggage & Mail ATSF 2109, 12/10; out prior to 1943.
29	Dynamometer	AT&SF 1910	51' 4wh	Retired, 2/63.
4999	Motion Picture	AC&F 1924	44' 4wh	Converted from Ft-M Flat car ATSF 97359, 3/31; reinstated as Ft-M Flat, 1943.
5000	Demonstrator	Pullman 1888	60' 4wh	Converted from Vestibuled Coach ATSF 500, 5/25.
2nd 5000	Diesel Instruction	Pullman 1926	79' 6wh	Converted from Cafe Observation ATSF 1508, 5/47; renumbered 80 Diesel Instruction, 1/22/73 Topeka.
5001	Power Plant Test	Pullman 1914	72' 6wh	Converted from ATSF 1338 Baggage-Buffet-Library, 9/39; to ATSF 5015 Dynamometer car, 9/61.
5002	Power Plant Test	SoCal 1898	41' 4wh	Converted from Superintendent's Business, ATSF 414, 1927; out prior to 1943.
2nd 5002	Physical Examination	Pullman 1916	82'5" 6wh	Converted from Pullman SAN RAMON, 5/50; retired, 2/60; sold 2/19/70.
5003	Physical Examination	Pullman 1910	72' 6wh	Converted from Diner AT&SF 1432, 7/29; destroyed by fire, 1935.
2nd 5003	Physical Examination	Pullman 1910	70' 6wh	Converted from Coach AT&SF 770, 8/35; to 199144, Bunk car, 11/53.
5004	Physical Examination	Pullman 1911	72' 6wh	Converted from Diner AT&SF 1433, 7/29; to 199145 Bunk car, 11/53.
5005	Air Brake Instruction	Pullman 1910	72' 6wh	Converted from Diner ATSF 1429, 8/30; retired, 2/60.
2nd 5005	Display	Pullman 5/30	70' 6wh	Originally Smoker ATSF 2952; to Chair-Smoker 3122, 1936; to Lounge-Dorm ATSF 1527, 5/48; to Display car 5005, Topeka, 6/67; renumbered AT&SF 87, Display car, 9/7/73, Topeka.
5006	Kansas Centennial	Pullman 6/16	82' 6wh	Originally 16-section MADRAS; to Tourist 4158; to ATSF Bunk 196474; converted to Exhibit 5006, 1954; retired, 1960; sold to W. Silver Co., El Paso, 12/23/70.
5007	Navajo Farm Instruction	Pullman 1911	72' 6wh	Converted from Diner ATSF 1440, 6/54; retired, 10/58.
5008	Diesel Simulator	Pullman 1947	85' 4wh	Converted from LW Chair-Lounge ATSF 3179, 5/69; renumbered 70 Diesel Simulator, 5/10/74, Topeka.
5009	Film	Pullman 1918	73½' 6wh	Converted from Business car ATSF 30, 3/69; renumbered 73 Film car, 1/16/73, Topeka.
5010	Air Brake Instruction	Pullman 1950	85' 4wh	Converted from LW Club-Lounge ATSF 1346, 8/72; renumbered 76 Instruction car, 1/4/74 Topeka.
5012	Rules Examiner	Pullman 1910	70' 6wh	Converted from Smoking car ATSF 586, 4/30; to Chair car AT&SF 1171, 1943.
5013	Rules Examiner	H&H 1882	51' 6wh	Converted from Superintendent's Business 2nd 412, 4/27: to 199181, Bunk car, 1/36; retired, 2/56.
5015	Test car (Dynamometer)	Pullman 1914	72' 6wh	Converted from Power Plant Test ATSF 5001, 9/61; renumbered AT&SF 83 Test car, 7/3/73 Topeka; to BNSF 83, Research & Test car, and named CANYON DIABLO.
5020*	Track Geometry	Pullman 1928	73½' 6wh	Converted from Business car ATSF 39, 10/73; renumbered AT&SF 85 Track Geometry car, 9/4/73, Topeka; to BNSF 85, Track Measurement. *Slated to be 5020, but new numbering scheme instituted prior to being out-shopped.

SANTA FE SPECIAL PURPOSE CARS

Car No.	*Type*	*Builder Date Built*	*Length/ Trucks*	*Notes & Disposition*
9000	Steam Generator			Converted from 1367 Loco tender, 3/9/43; reno. 1, 8/67; retired, 6/13/68.
9001	Steam Generator			Converted from 1360 Loco tender, 4/1/41; reno. 2, 8/67; retired, 9/27/67.
9002	Steam Generator			Converted from 1361 Loco tender, 6/17/43; reno. 3, 8/67; retired, 9/27/67.
9003	Steam Generator			Converted from 1364 Loco tender, 8/43; reno. 4, 8/67; retired, 6/13/68.
9004	Steam Generator			Converted from 1357 Loco tender, 12/28/43; reno. 5, 8/67; retired, 6/13/68.
130	Steam Generator	ATSF/Pull 1956	63'11" 4wh	Converted from Baggage 3707, 11/28/62, numbered 9005; renumbered 6, 1/68; reno. 130, 3/12/70.
131	Steam Generator	ATSF/Pull 1965	73'11" 4wh	Converted from Baggage 3927, 5/70.
132	Steam Generator	ATSF/Pull 1965	73'11" 4wh	Converted from Baggage 3920, 5/70.
133	Steam Generator	ATSF/Pull 1965	73'11" 4wh	Converted from Baggage 3922, 5/70.
134	Steam Generator	ATSF/Pull 1965	73'11" 4wh	Converted from Baggage 3916, 5/70.
135	Steam Generator	ATSF/Pull 1965	73'11" 4wh	Converted from Baggage 3936, 5/70.
136	Steam Generator	ATSF/Pull 1965	73'11" 4wh	Converted from Baggage 3933, 5/70; sold to Auto-Train 136, 10/15/71.
137	Steam Generator	ATSF/Pull 1965	73'11" 4wh	Converted from Baggage 3932, 6/70; sold to Auto-Train 137, 10/15/71.
138	Steam Generator	ATSF/Pull 1965	73'11" 4wh	Converted from Baggage 3926, 6/70; sold to Auto-Train 138, 10/15/71.
139	Steam Generator	ATSF/Pull 1965	73'11" 4wh	Converted from Baggage 3918, 7/70.
140	Steam Generator	ATSF/Pull 1965	73'11" 4wh	Converted from Baggage 3931, 7/70; renumbered 2nd 138 Steam generator, 8/74.

AIR BRAKE INSTRUCTION CARS 99905-99906

In 1900, Santa Fe set up AIR BRAKE INSTRUCTION car 99906 from an old 2500 series Coach, followed by car 99905 in 1903, converted from a 900 series Postal car. The General Air Brake Instructor was in charge of the cars, the interior of which were similar.

Each made two trips a year over its assigned territory. Car 99905 covered from Chicago to Dodge City, including the Pecos Valley and Gulf Lines. Car 99906 covered from La Junta west including the Coast Lines.

The cars were furnished with a 9½-inch air pump, which was operated by air pressure. Power was furnished from yard plants at important division points. Underneath the car was a reservoir to which the yard line pipes ran, which was used for storage, and a pipe from this ran to the air pumps, the discharge from the air pumps going to another reservoir, which was used as a main reservoir.

The instruction room was 38 feet long with a seating capacity of 30. The equipment consisted of 19 Westinghouse 8x12-inch freight brake cylinders arranged vertically along each side of the car and one New York 8x12-inch freight brake cylinders; two Westinghouse 10x12-inch passenger brake cylinders and two Westinghouse 10x10-inch driver brake cylinders. The equipment in these cars was developed to represent a train line equal to 50 cars, and there was a standard 26x36-inch reservoir which was used as an auxiliary to the train line.

Distributed along the side of the car below the equipment were the sectionals of the working air pump showing a tandem with the pump, the triple valve operating the 10x12-inch passenger brake cylinder, having a tandem sectional triple attached. It was also equipped with a slack adjuster which was cut in sections to show its operation. Overhead was sufficient pipe to represent six cars of air signals. The signal cord handles were arranged conveniently for the instructor to operate at the demonstration table. This table had on its front the Westinghouse G-6 brake valve. Over the operating table was a rack containing charts of the different equipment. Under the table on which the sectionals were arranged was a cutaway of standard freight equipment with triple valve, bleeder and retainer, also sectionalized. The defective triples were mixed in with the operating triples and could be either operated or cut out.

On the instructor's office desk was a duplex gauge showing the train line and brake cylinder pressure on the train when running attached to two Crosby recording gauges which produced a chart that was forwarded to the master mechanic of each division for review.

Instruction cars 99905 and 99906 were subsequently renumbered 199905 and 199906 in 1912, in keeping with the renumbering of all maintenance-of-way equipment. The two cars were again renumbered in 1918 to 26 and 27. The 26 was replaced in 1923 and the 27 in 1926, both by newer, larger cars numbered 2nd 26 and 27.

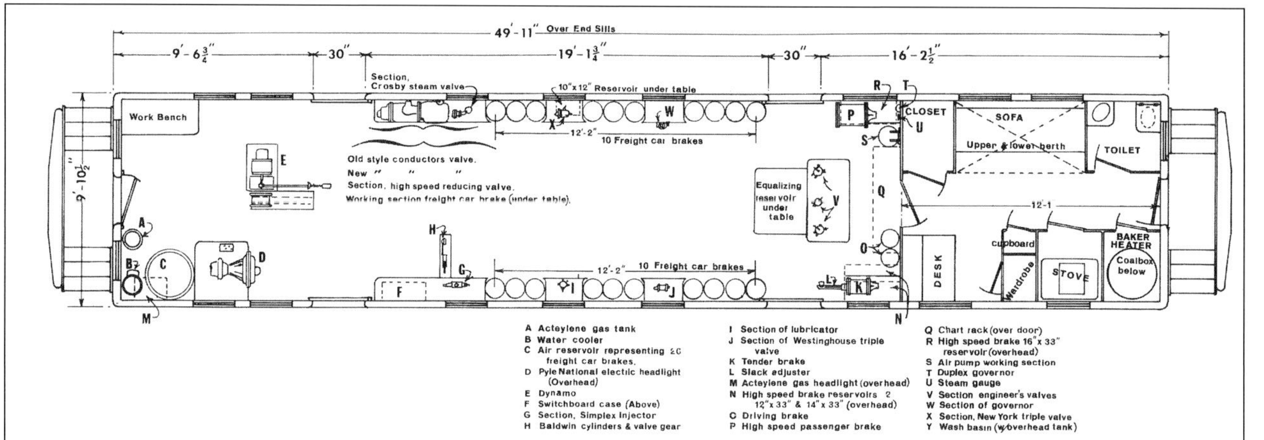

Santa Fe's second AIR BRAKE INSTRUCTION car, 99906, complete with gold striping, stands at one of its locations ready for classroom duty about 1907. The flat boards (one behind the brake staff) were hinged to lower, forming a full width platform in the "down" position. This car was renumbered 1st 27. *-Santa Fe photo, Frank Ellington collection*

20 A.T.&S.F. TRACK RECORDING CAR

AT&SF 20, originally built by Barney & Smith in 1886 as Gulf, Colorado & Santa Fe Business car 217, then GC&SF 12. The 60-foot six wheel car was renumbered AT&SF 20, Business, in 1910. In July 1927 it was taken into Topeka Shops and fitted as a TRACK RECORDING CAR. The view above shows car 20 at Topeka just released from the shop. At right, is an interior view of TRACK RECORDING CAR, 20, showing the switch board, gyroscope, master clock, gauges, and chronograph. *-Above, Santa Fe photo, John B. McCall collection; right, Richard Scholz collection*

Developed by Major J. deN. Macomb, office engineer, in collaboration with E. E. Chapman, engineer of tests, the track recorder consisted of mechanisms to record the following information: (1) Low joint recorder, (2) Track gauge recorder, (3) Lateral movement of truck and car body recorder. (4) Car body inclination and cross level of track, controlled by a gyroscope, (5) Record of time intervals controlled by a clock (to determine speed), and (6) Location of landmarks recorded by push button. All this information was recorded by pens on paper, passing over a platen or table.

The Low Joint Mechanism was fitted to the six wheel truck under the observation end of the car in which the front and rear axles were rigidly connected by a bar, while the center axle was permitted a free vertical movement. Any movement below the common level of the outer wheels was recorded. In addition, mechanical counters, recorded the number of low joints of each interval on both sides of the track.

A continuous record of track gauge was made by means of two wheels six inches in diameter pressing against the gauge side of the head of each rail. When the gauge recording device was not in service the whole mechanism was lifted upward about four inches by releasing the air on the special piston. By means of the gauge mechanism a record of the lateral movement of the equipment between the rails with respect to the right rail was also secured. The lateral movement mechanism included a gyroscope which indicated the cross level.

A clock was provided with two small pens and two electric contact levers mounted within it in such a way that an electric contact is made at desired intervals, reaching from ten to thirty seconds, which registered time, speed and location. The speed of the train at any instant could be read directly by applying a scale to the record. The pen, which recorded the time intervals was also utilized to show the various landmarks such as mile posts, stations, bridges, switches or road crossings. The circuit which operated the pen for this purpose was closed by means of a push button manipulated by an observer stationed on the rear platform of the car at the instant the landmark was passed. The observer on the rear platform remained in communication with the operator at the desk by means of telephone.

The record paper was drawn across the table at a rate of 13.2 in. per mile of car run regardless of the speed of the car. This afforded a convenient scale which corresponded with standard Santa Fe engineering department profile paper. Upon the scale a 33 ft. rail length was approximately 1/12 in. The paper was driven by means of a flexible cable drive from the axle of the truck and reduction of speed was made by the use of gears. *-From Railway Age-Railway Engineering and Maintenance, March, 1927; Vol. 82, No. 12; Pages 767- 768.*

21 A.T.&S.F. Radio Test Car

Originally built by Barney & Smith in 1890 as Diner AT&SF 1407, the 65-foot, six wheel car was converted to Business car AT&SF 21 in 1910. Car 21 was fitted out with special equipment right after World War II and became Santa Fe's system electronics laboratory, or Radio Test Car. Car 21, seen at Topeka, Kansas, in May 1955, was used to usher in the radio communications era during the mid and late-1940s. Over the years it was in service, it was seen with various devices and antennae fitted to its roof. That radio communications became a major link between trains and dispatchers, dispatcher and mobile units of all kind is an established fact, but in February and March 1947, car 21 even conducted a series of two-way V.H.F. radio tests involving a twin-engine, Santa Fe Skyways Douglas airfreighter. Radio Test car 21 was sold to the Stockton Terminal & Eastern Railroad in 1958 and was converted into stationary offices. *-F.O. Kelley photo, John B. McCall collection*

27 A.T.&S.F. (2nd) Instruction

2nd 27, one of several Air Brake Instruction cars once rostered by Santa Fe, was converted in June 1926 from 70-foot Smoker ATSF 567, built by Pullman in 1912. The car, seen here at Topeka on January 12, 1941, served along side Air Brake Instruction car 5005 and was retired twenty years later. *-Ray Hilner photo, John B. McCall collection*

29 A.T.&S.F. Dynamometer Car

As early as 1905, Santa Fe was testing the proficiency of their locomotives with leased Dynamometer cars. A Scranton Dynamometer car was used in testing from Florence to Winfield, Kansas, on November 22, 1905 and a Westinghouse Dynamometer car tested from Holliday to Olathe, Kansas, on December 27, 1905. By 1910, Santa Fe had constructed their own Dynamometer car at the Topeka Shops and numbered it 29. Over the years, the Dynamometer car made test runs with just about every class of steam locomotive on the system in service after 1910, and went on to test diesel locomotives as well ending it's career in February 1963. Here we see both sides of this unusual car at Topeka, April 4, 1951. *-F.O. Kelley photos, John B. McCall collection*

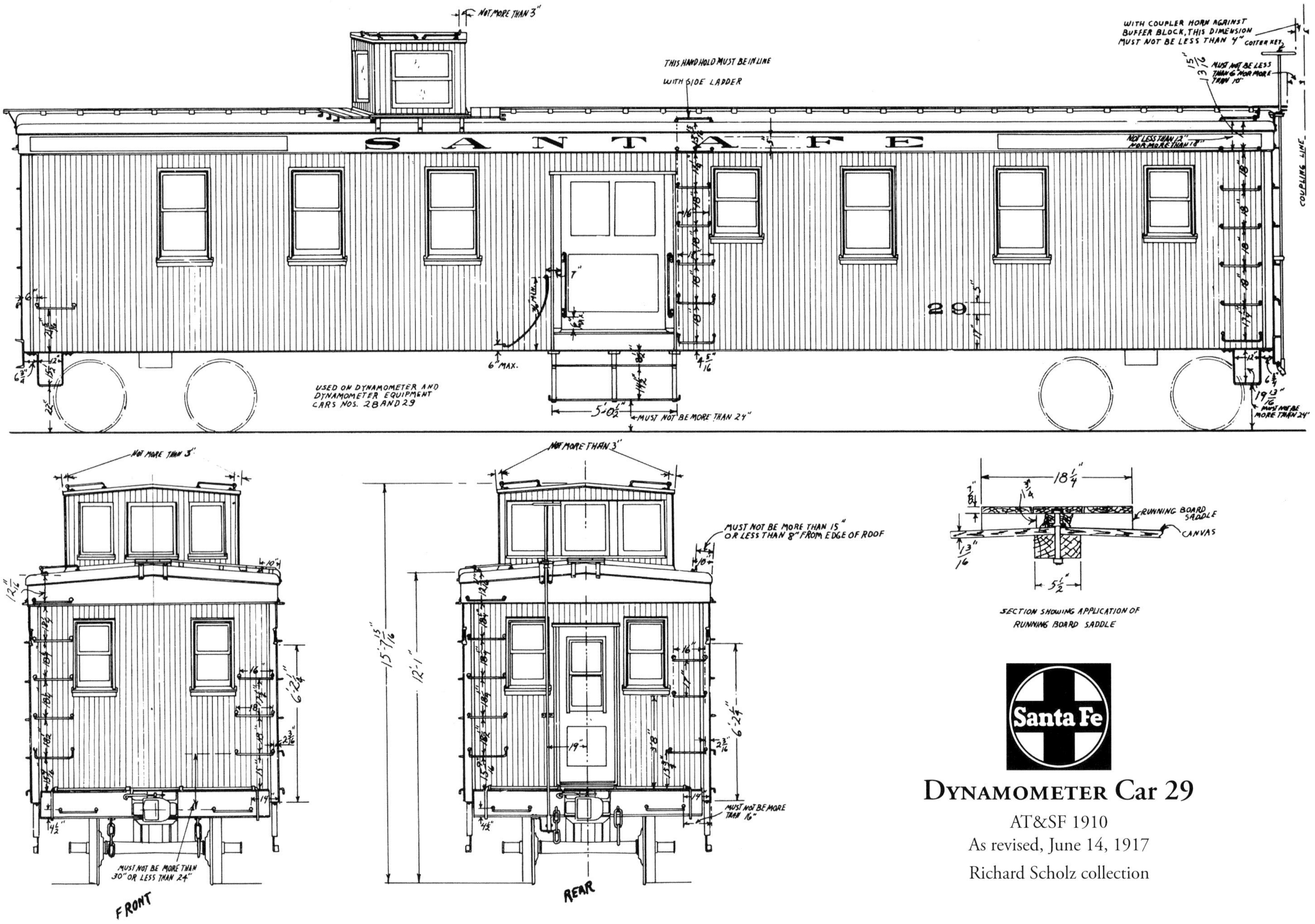

Dynamometer Car 29

AT&SF 1910

As revised, June 14, 1917

Richard Scholz collection

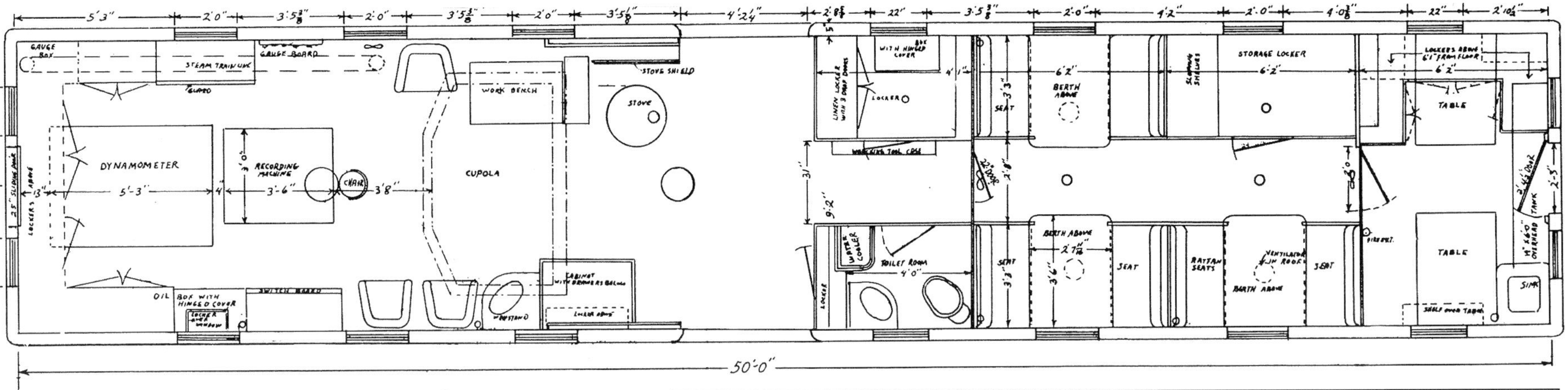

Dynamometer Car 29

AT&SF 1910

From drawing 18-C 469, as revised, June 24, 1944

Drawn by Frank Ellington

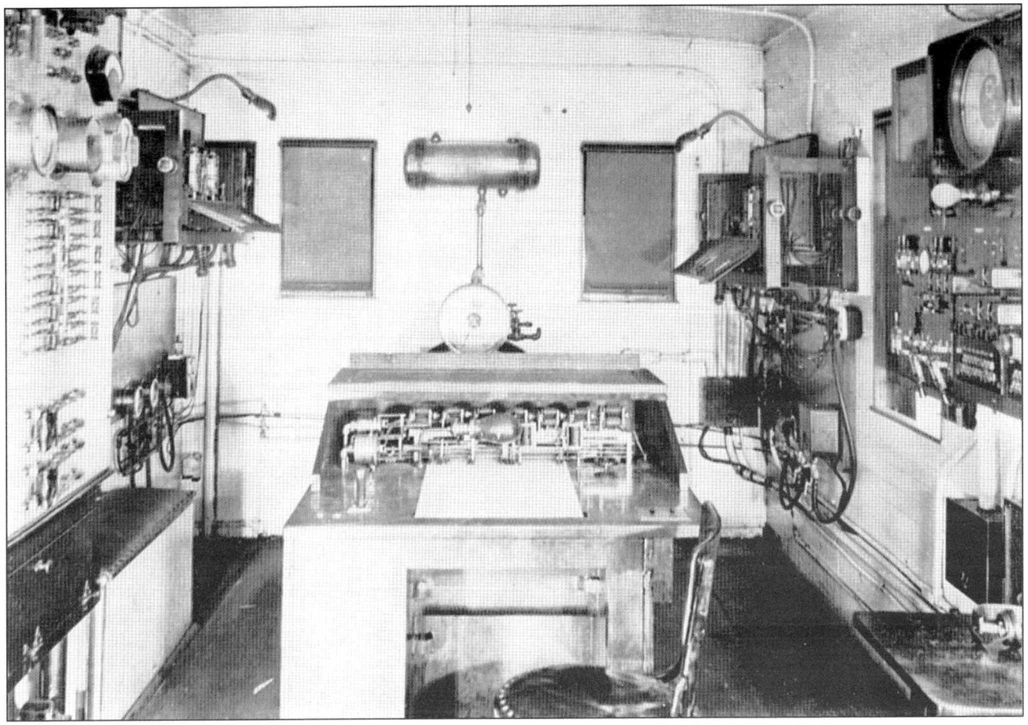

An interior view of car 29 looking forward. Later there will be a door cut between the two windows. Notice the switchboard to the left, the chronograph drawbar pull gauge, dynamometer fluid reservoir in the center, and special electrical instruments on the wall used in air brake tests for the A.A.R. *-Richard Scholz collection*

Left: Car 29 in storage at Topeka on June 29, 1962. During its first years of operation there was no door in this end of the car. Note the multiple unit plugs for diesel electric locomotive functions. *-John B. McCall collection*

Dynamometer car 29 and Dynamometer Equipment car No. 28 are seen here at San Diego in 1920 being used in a test of Pacific 1276 while in actual service, ready to leave the San Diego depot with an eleven car consist. The housing on the front of the locomotive was to protect Test Department observers from the elements as the tests were conducted. *-Richard Scholz collection*

Forty-eight year old Dynamometer car 29 is seen in service on November 16, 1948, at San Bernardino on the First District of the Los Angeles Division cut in behind F3s on train No. 24, the *Grand Canyon*. The car would later receive roller bearings, but though modernized on the inside to be able to monitor diesel performance, the exterior, including the staff brake, remained essentially the same for its entire career. *-Jack Whitmeyer photo, SFRH&MS Collection*

5000 A.T.&S.F. Diesel Electric Instruction

2nd 5000, Diesel Electric Instruction car, was converted from 1926 Pullman-built 79-foot Cafe Observation ATSF 1508, in May 1947. A little over a year later, on November 12, 1948, the car is at San Bernardino. The car was renumbered 80, Diesel Instruction, at Topeka on January 22, 1973. *-Jack Whitmeyer, SFRH&MS Collection*

A couple of year later, at Chanute, Kansas, in 1950, the three side windows behind the sliding door had been blanked over. Toward the end of its service life, all of the side windows will have been plated over. *-Wally Abbey*

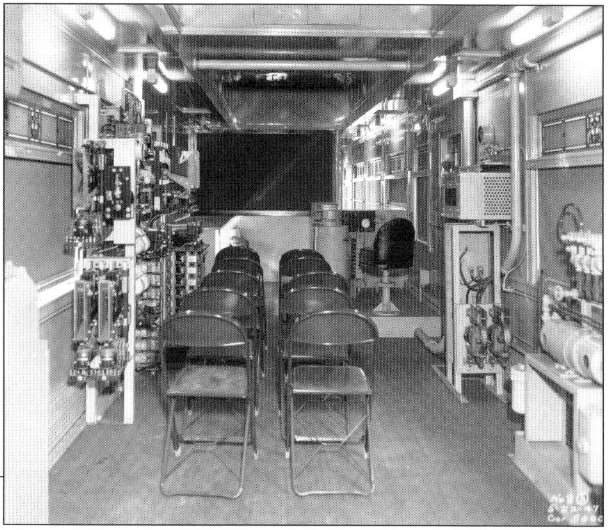

Completed in December 1947, Diesel Electric Instruction car 5000 was designed to provide practical training for shop personel and enginemen in the latest, up-to-the-minute methods of operating and maintaining Santa Fe's rapidly growing diesel fleet, the largest in the world at that time. The car was designed by Santa Fe's mechanical department as a classroom on wheels. While it had seats and a blackboard, it was more like a scientist's laboratory, as it contained all of the working parts of a diesel engine with the exception of the generator and traction motors. One of the outstanding features was a cutaway of a two-cycle diesel engine. All of the working parts of the engine were exposed so that even the beginner could understand its operation. Other installations include a steam generator, electrical contact racks, control station and even the familiar diesel horn. When completed, it departed Chicago in the latter part of June 1947 for the first lap of a year-long journey during which stops were made at every important terminal on the system. In addition to using the equipment on board to describe and demonstrate the operation and repair of diesel engines, Supervisor of Diesel Instruction Glenn E. Rodgers augmented his instruction with chalk talks, color slides and movies. That's Mr. Rodgers at the projector, who traveled with the car for its first tour. The above view was taken at Marceline. Missouri in June 1947, while the two interiors at right show the completed car at Topeka on May 22, 1947. *-Santa Fe photos, Railroad & Heritage Museum Collection, Temple, Texas*

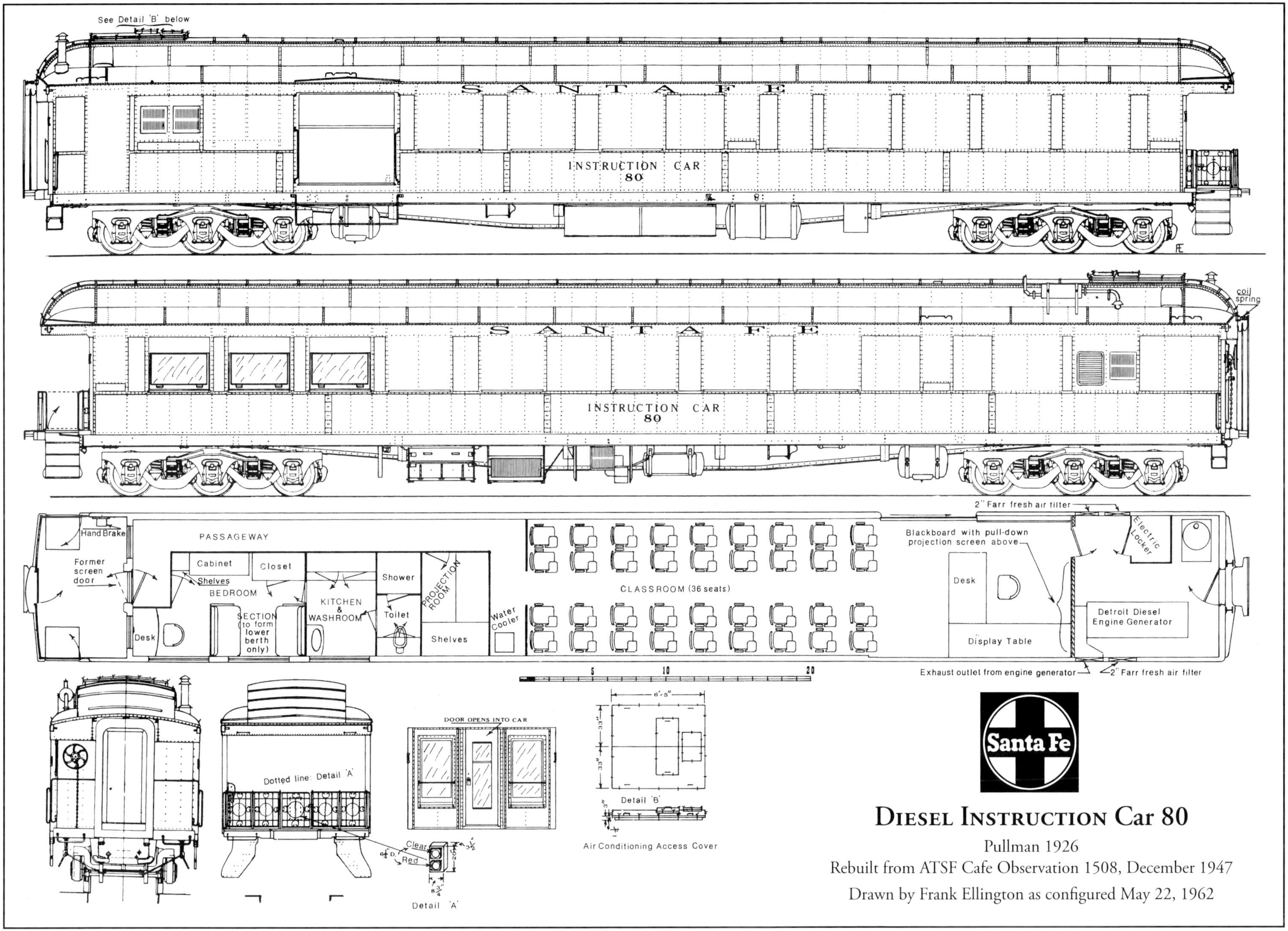

Diesel Instruction Car 80

Pullman 1926

Rebuilt from ATSF Cafe Observation 1508, December 1947

Drawn by Frank Ellington as configured May 22, 1962

Car 5000 was revised on the inside continuously as diesel technology changed. In the view, left, taken in 1960 at Barstow, California, all side windows had been plated over, except for the living quarters, but the car still carried its very ornate observation railing. Then in 1962, the interior of the car was completely remodeled with virtually all of the equipment and controls removed to make way for an expanded classroom area and repurposed as an INSTRUCTION CAR. By August 27,1973, at Topeka, Kansas, below, the car had been renumbered 80 and SKF roller bearings had been added. *-Left, Seeth "Corky" Thompson photo, SFRH&MS Collection; below, Joe McMillan*

5002 A.T.&S.F. Power Plant Test Car

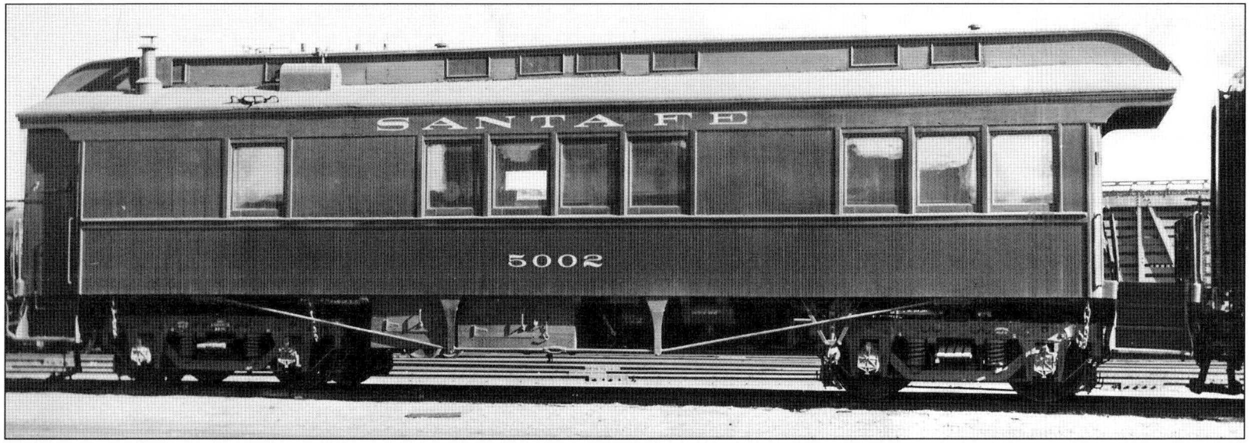

The system's original Power Plant Test, 1st 5002, car stands idle at Topeka, February 6, 1938, a couple of years prior to retirement. Originally built by the Southern California Railway in 1898 as Officers car 97, it was renumbered AT&SF 414, Superintendent's Business car, in June 1904. The 41-foot car was converted to AT&SF 5002, Power Plant Test car in 1927. As superintendent's business car 414 it received an electric generator, and storage batteries for electric lights during the early 1920s, but they were removed when converted, and the battery boxes contain equipment here. At right is a view of the instrument room taken in 1927. The car was off the roster prior to 1940. *-Above, Ray Hilner; right, Richard Scholz collection*

In late 1927, Santa Fe placed in service a Power Plant Test car which, with its testing equipment, was designed to provide a uniform measure of operation for the eighty major power plants, exclusive of small pumping and heating stations, located on the system. Data collected by means of this car enabled the management to obtain an accurate knowledge of power plant costs and the visits of the car served a secondary useful purpose of disseminating information concerning desirable practices among the various power-plant operators.

The geographical location of the plants in respect to the fuel supply resulted in 30 of these plants using coal as fuel, 40 using oil and 10 using natural gas. The larger plants furnished both alternating and direct current, compressed air, water and steam to serve the local facilities. Other plants had no electrical generation equipment while the smaller stations pumped water and furnished steam only.

Fueling the eighty power plants cost $1,323,000 during 1928 and their operation incurred an additional charge of $2,039,000. It was Santa Fe policy to make tests to check the performance of power plants with an inspector visiting the various plants to check upon operation, to standardize methods and equipment and to correct wasteful practices. The cost per pound of steam, the distribution of the steam generated, the relative cost of fuels as affected by the different handling charges and accessories required, called for more careful investigation than could be obtained from the ordinary plant records or inspections.

It was in recognition of all this that the Power Plant Test car was developed. The car provided space for test equipment and also office and living quarters for the test crew. The instrument room occupied the largest portion of the car and in it were located the racks for the instruments, a motor generator set, Baker heater, lockers, a work bench, and miscellaneous cabinets built to meet the requirements of the special apparatus. Each instrument was fixed in place and the portable instruments were located so as to entail the minimum loss of time in placing them in operation. The arrangement of the observation end of the Business car was well suited for an office without changes other than the addition of a small desk and folding table.

Further space for miscellaneous equipment and fittings was provided by suspending equipment boxes beneath the car. Suitable connectors were located at each end of the car for plugging into the regular station or powerhouse lighting circuit or into an adjacent car while enroute. Oil lamps or flashlights were used when electrical current was not available.

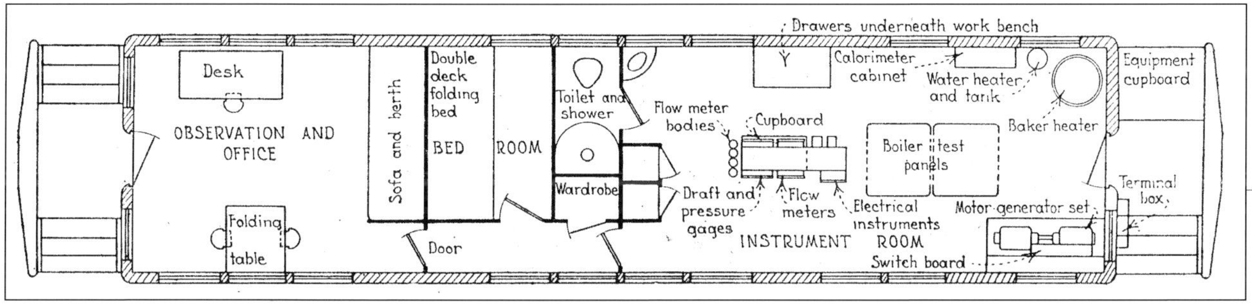

PHYSICAL EXAMINATION CARS

Car 2nd 5002, seen at Topeka, April 22, 1951. Built by Pullman in January 1917 as a 10 Section-Lounge-observation SAN RAMON, this car spent many years assigned to Burlington Route trains. The Pullman Company sold the car in 1947 to a private party, and then it was resold to the Santa Fe in 1950, which promptly converted it to a Physical Examination Car and renumbered it 5002. This was the last Physical Examination car on the system. *-Ray Hilner*

All three of the early Physical Examination cars, 5003, 5004 and 2nd 5003, were converted from wooden cars with steel underframes. EYE AND EAR EXAMINATION car 5004, seen here October 5, 1929, coupled to PHYSICAL EXAMINATION car 5003, was converted from a Pullman 1911-built Dining car as a companion to the 5003. *-Santa Fe photo, Frank Ellington collection*

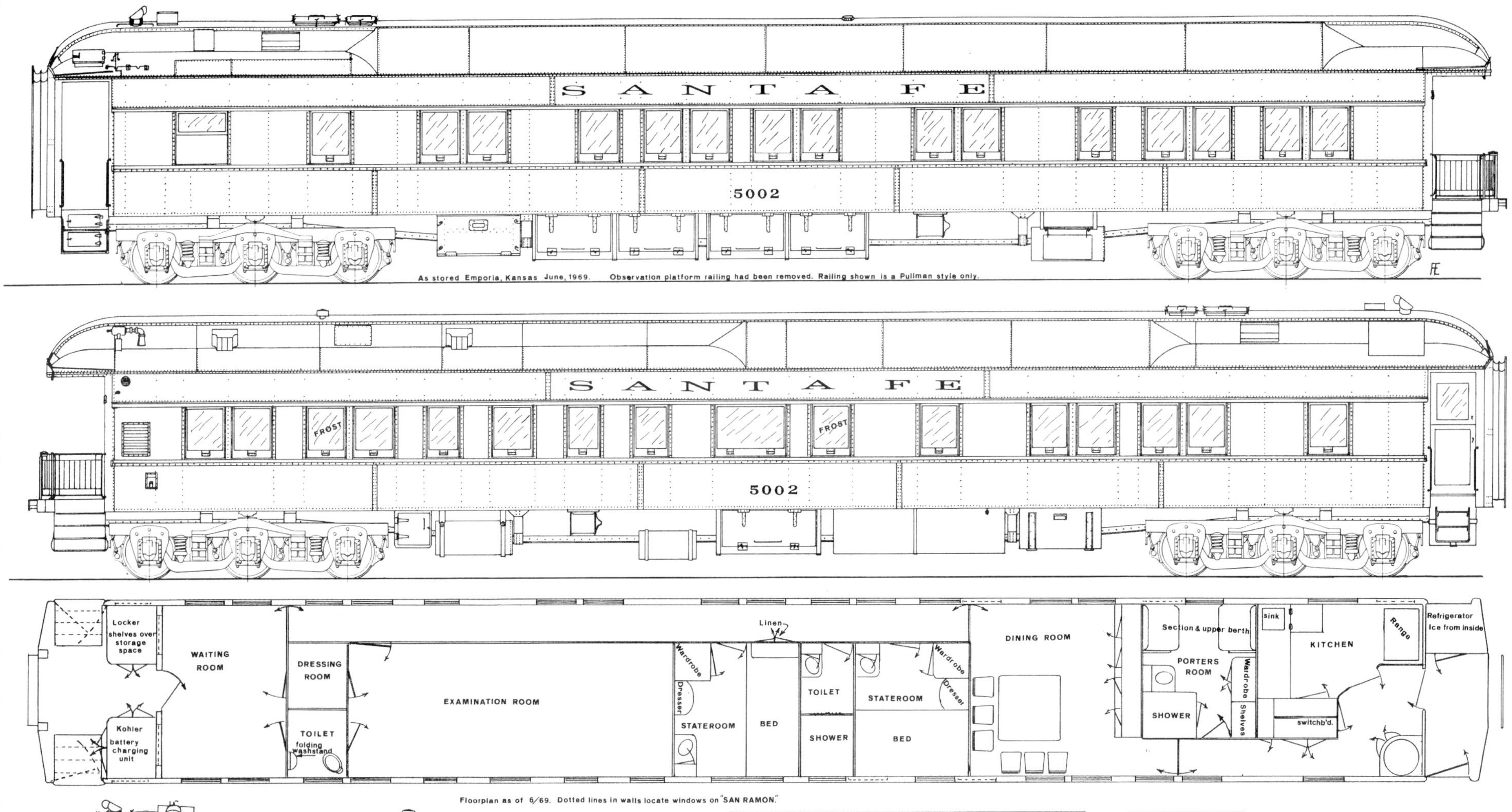

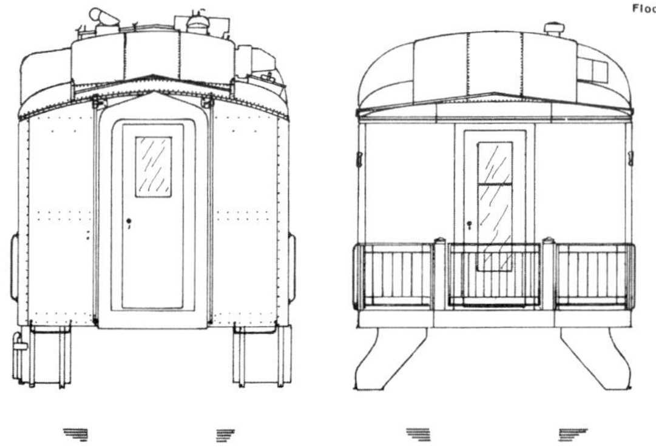

Physical Examination Car 2nd 5002

Pullman 1917

Rebuilt from Pullman 10-section-lounge San Ramon, 1950

Drawn by Frank Ellington as stored at Emporia, Kansas, June 1969

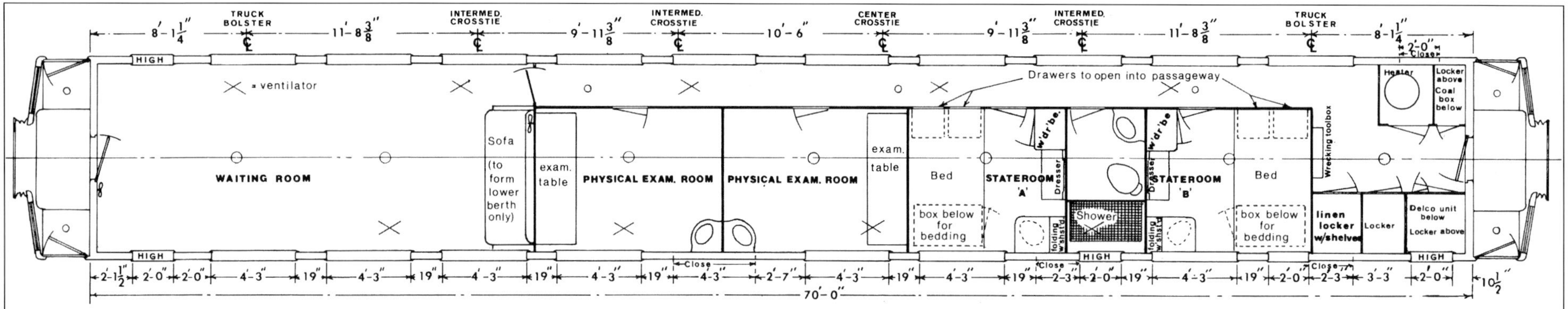

Physical Examination Car 5003

(Rebuilt from Coach 770)

August 1935

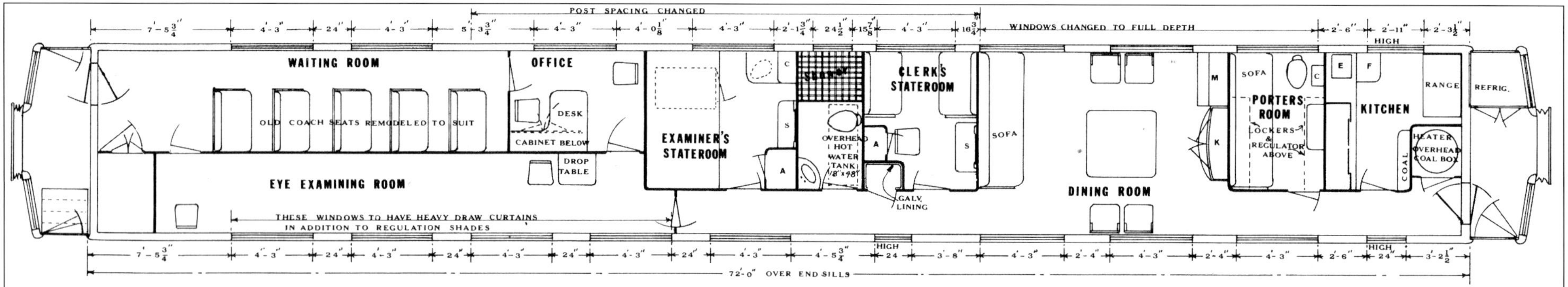

Physical Examination Car 5004

(Rebuilt from Diner 1433)

July 31, 1928

EYE AND EAR EXAMINATION car 5004 (ex- Diner 1433) served the Santa Fe in that capacity for almost a quarter of a century before being converted to a Maintenance-of-Way Bunk car in 1953 and renumbered 199145. It is seen here at Strong City, Kansas, February 22, 1962, looking a little tattered. *-John B. McCall*

5005 A.T.&S.F. Air Brake Instruction

Taking the place of Air Brake Instruction 2nd 26. which had been converted from an old Parlor-Observation car, the instruction room in Air Brake Instruction 5005 was very similar to those of all the older air brake instruction cars. The instruction room had a compartment forty feet long, with ten Westinghouse freight brake cylinders arranged vertically along either side of the car. Amidst a maze of piping, tanks, and gauges, Santa Fe instructed its employees in the proper usage of modern air brake equipment by practical demonstration. Distributed along the side of the car below the equipment were the sectionals of the working air pump. It was also equipped with a slack adjuster which was cut in sections to show its operation. There was no boiler in the car and power and air were furnished from yard plants at important division points. The General Air Brake Instructor was in charge of this car. Note the open upper deck transoms. *-Santa Fe photo, Railroad & Heritage Museum Collection, Temple, Texas*

Less than four months after its release from Topeka Shops, Air Brake Instruction 5005 poses for its portrait, right, on December 26, 1930. Converted from 1910 Pullman-built Diner ATSF 1429, the 72-foot car served the Santa Fe until February 1960. During its conversion from the Diner, the kitchen end was reconstructed with an observation platform. *-Santa Fe photo, Frank Ellington collection*

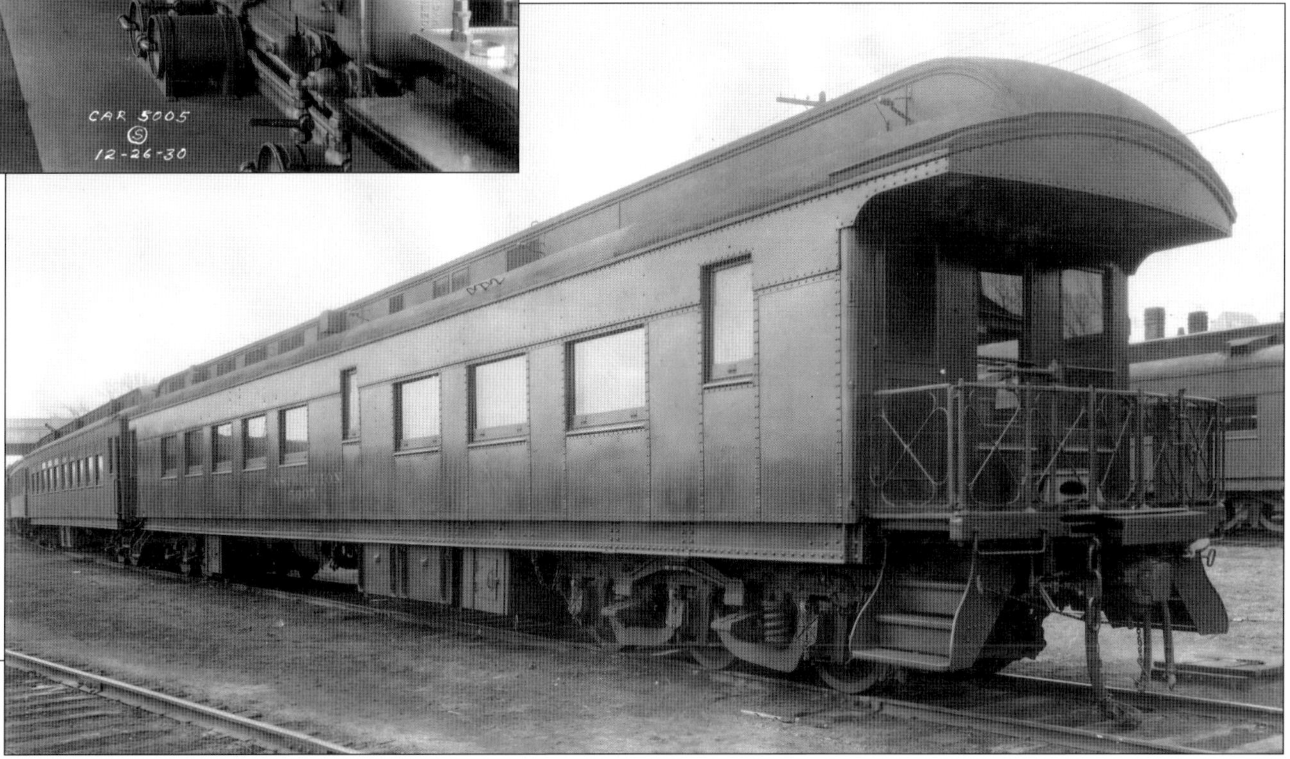

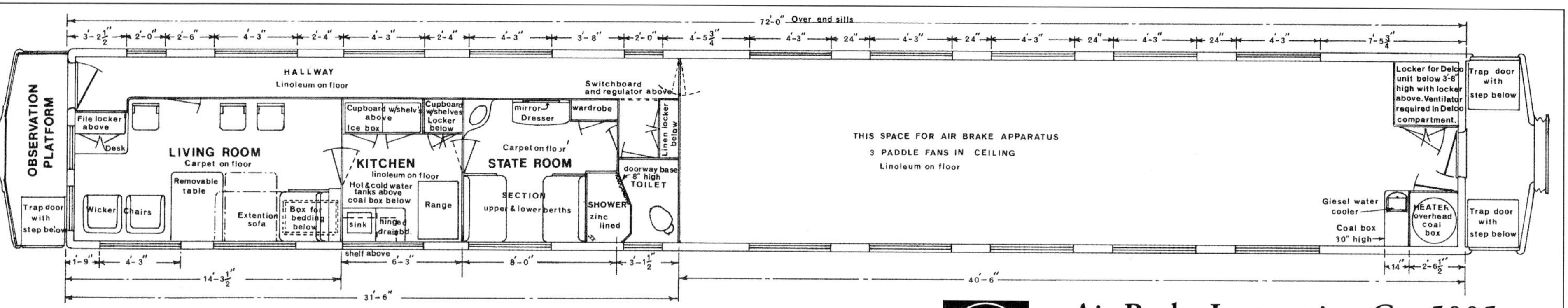

Air Brake Instruction Car 5005

(Rebuilt from Diner 1429)
August, 1930

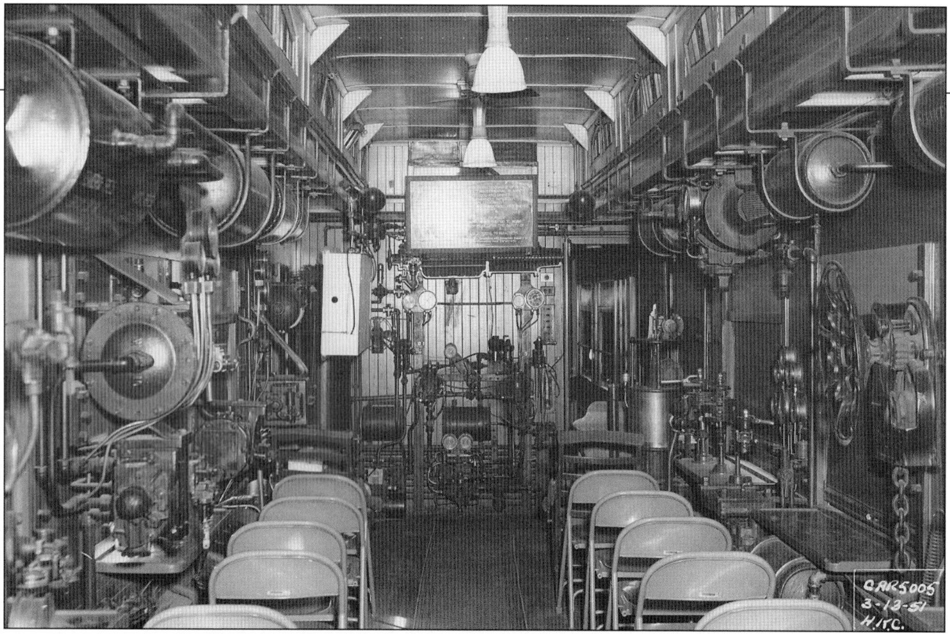

During the 1950s Air Brake Instruction 5005, seen here at Topeka, Kansas, March 12, 1951, was air conditioned, making some modifications necessary. The air conditioning unit was located in the front, left side. The vestibule was sealed on this side, the first window removed and replaced the small window just ahead of the observation mid-car behind the former observation lounge. The interior apparatus was continually modified and upgraded as changes were made in new equipment arriving on the property. *-Left, Santa Fe photo, Railroad & Heritage Museum Collection, Temple, Texas; above, Ray Hilner, John B. McCall collection*

The opposite side of Air Brake Instruction 5005, at Topeka, on August 14, 1938. During more than twenty years of service as an Air Brake Instruction car the 5005 showed very little change except the Pullman green paint gave way to Santa Fe's coach green and the gold lettering was replaced with yellow. In later years, the roof was painted in reflective silver paint. *-Ray Hilner, Frank Ellington collection*

5005 A.T.&S.F. (2nd) Exhibition

Smoking car ATSF 2952, built by Pullman in May 1930, was converted to a Chair-Smoker in 1936 and renumbered 3122. Topeka Shops converted the 3122 to a Lounge-Dormitory car in May 1948 and renumbered it 1527. This car, seen at Los Angeles about 1960, was destined to be converted to an Exhibition car, 2nd 5005. *-Frank Ellington collection*

The 1527 was converted at the Topeka Shops in June 1967 to Exhibition car 2nd 5005 for the Chisholm Trail Centennial celebration. Lettered for the Chisholm Trail Centennial Museum, it was painted beige with silver roof and trucks. The colorful flags of Kansas and Oklahoma with their blue background and multicolored centers, and the red, white and blue "Lone Star" flag of Texas graced the flank. "Santa Fe" was in black, and lower lettering and numbers, were in red. The car is seen in its new configuration at Topeka on October 6, 1968. *-Two photos, Joe McMillan*

Car 2nd 5005, seen at Topeka, left and below in 1971, was subsequently reoutfitted and repainted dark blue and light gray for the Panorama of Progress display in 1970. One set of trucks were gray while the set on the blue end were black. The car number and the lettering "Panorama of Progress" were in black while the herald and the lettering "Always on the move toward a better way" was in yellow. The vertical ring around the herald was yellow, the horizontal was white and the two on angles were red and light blue respectively. *-Two photos, Frank Ellington*

Finally 2nd 5005 received the red, white and blue scheme in 1972 and became part of the spectacular red, white, and blue "Spirit of '76" exhibit train which was launched November 11, 1972, from the Santa Fe depot at Richmond, California bound for a cross-country tour. The exhibit was visited by an impressive 36,570 visitors at Santa Fe system points alone. The colorful train consisted of Seaboard Coast Line U36B, Santa Fe system Generator car 133, Exhibit car 5005, and a Frisco caboose. The diesel and caboose were both numbered 1776. The car was renumbered AT&SF 87 Display car at Topeka September 7, 1973, where it is seen one month later. *-Frank Ellington*

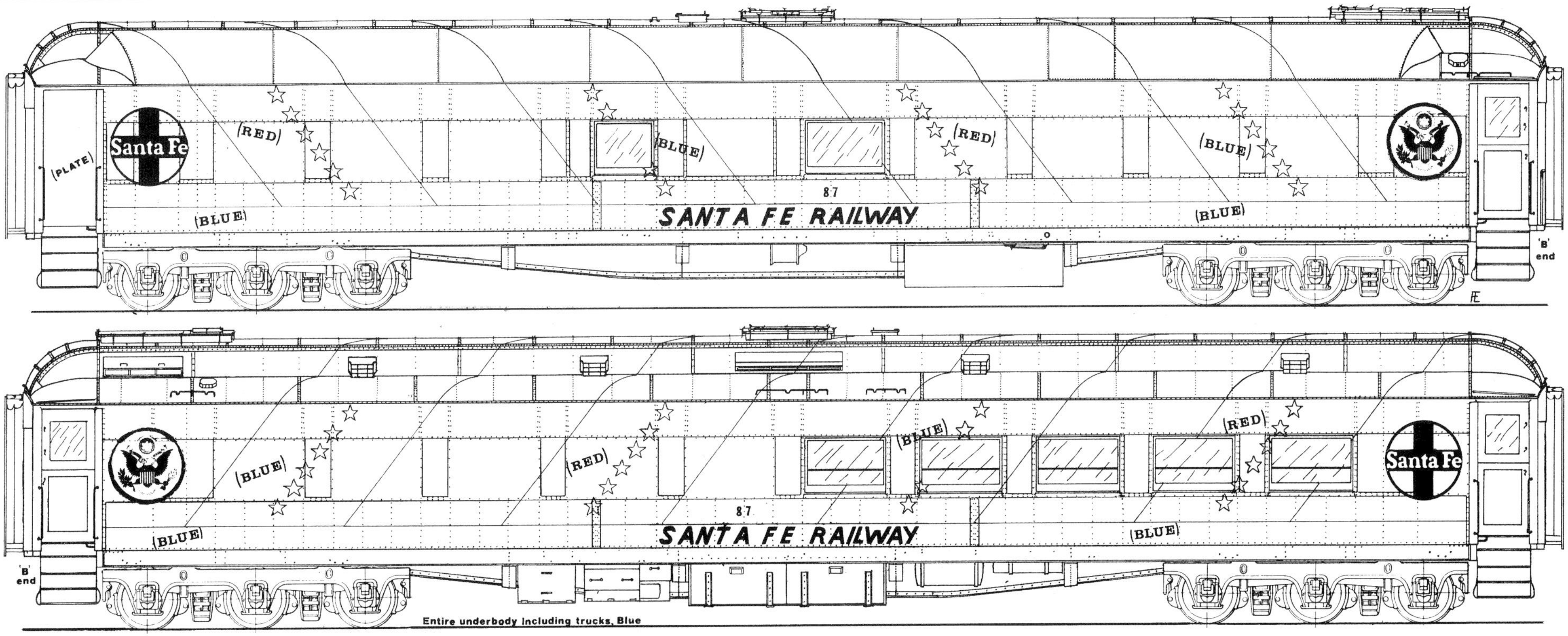

Exhibition Car 2nd 5005

Pullman 1930

Rebuilt from Lounge-Dormitory 1527, 1967

Drawn by Frank Ellington as stored at Topeka, February 1975

ATSF Exhibit car 87 at Topeka, in August 1975. *-Louis Marre photo, John B. McCall collection*

5006 A.T.&S.F. Kansas Centennial

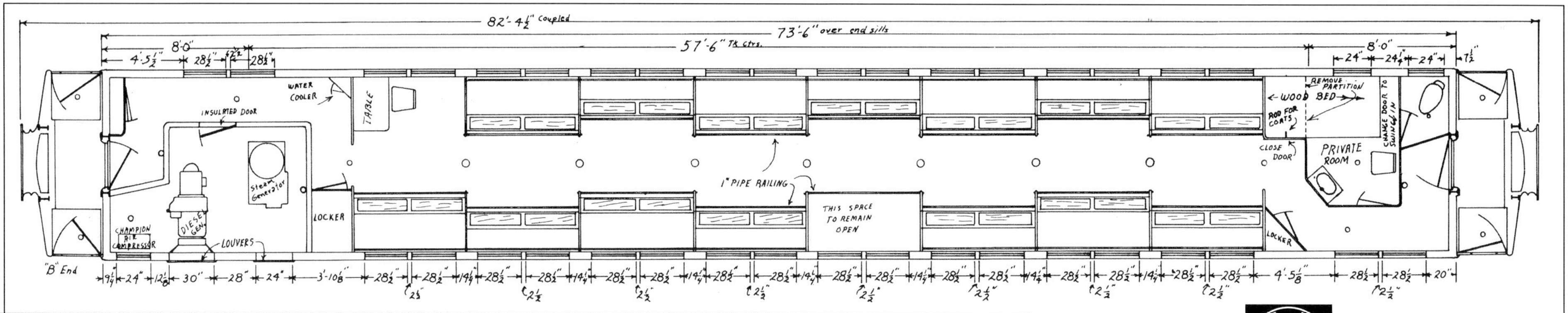

Kansas Centennial Car 5006

Originally 16-section Pullman Madras, 1916

(Rebuilt from Bunk 196474, 1954)

Left: The Pullman Company built 25 16-section cars in May and June 1916 to plan 2412F in lot 4386. One of these cars was named Madras which was converted by the Pullman Company to Tourist 4158 in July 1935. Tourist 4158 was sold to the Santa Fe in September 1953 and converted to ATSF 196474, Bunk car. Four months later the 196474 was selected for the Kansas Territorial Centennial Exhibit car. Its interior was stripped and fitted with exhibit cases. It was painted in the two-tone gray scheme with a black roof, renumbered 5006 and lettered Kansas Territorial Centennial Exhibit. *-John B. McCall collection*

Car 5006, Kansas Territorial Centennial Exhibit, is seen in dead storage at Topeka, right, on April 10, 1966. The car was retired in 1960 and sold to scrapper W. Silver Co. of El Paso, Texas, on December 23, 1970. *-John B. McCall*

5008 A.T.&S.F. Locomotive Simulator

Introduced in September 1969, the $750,000 Santa Fe Locomotive Simulator, car 5008, was the first such training device to appear on any U.S. railroad, and was built into former 1947-built Pullman 85-foot lightweight Chair-Lounge ATSF 3179. Seen here at Shopton, Iowa, in 1971, it was renumbered 70, Locomotive Simulator, at Topeka on May 10, 1974. *-Frank Ellington*

Inside the simulator the student experienced all the visual, audio and motion sensations normally encountered while operating a train over the railroad. The operating controls in the cab were authentic, and the rear screen projection unit, developed by Link Systems, featured a "collimating lens" located just in front of the cab window. Instructors monitored the reactions of the trainee, and made suggestions for improvement needed. All recording scenes which appeared in the simulator's visual system film were shot using Camera car 5009. *-Two photos, Santa Fe Collection, Kansas State Historical Society*

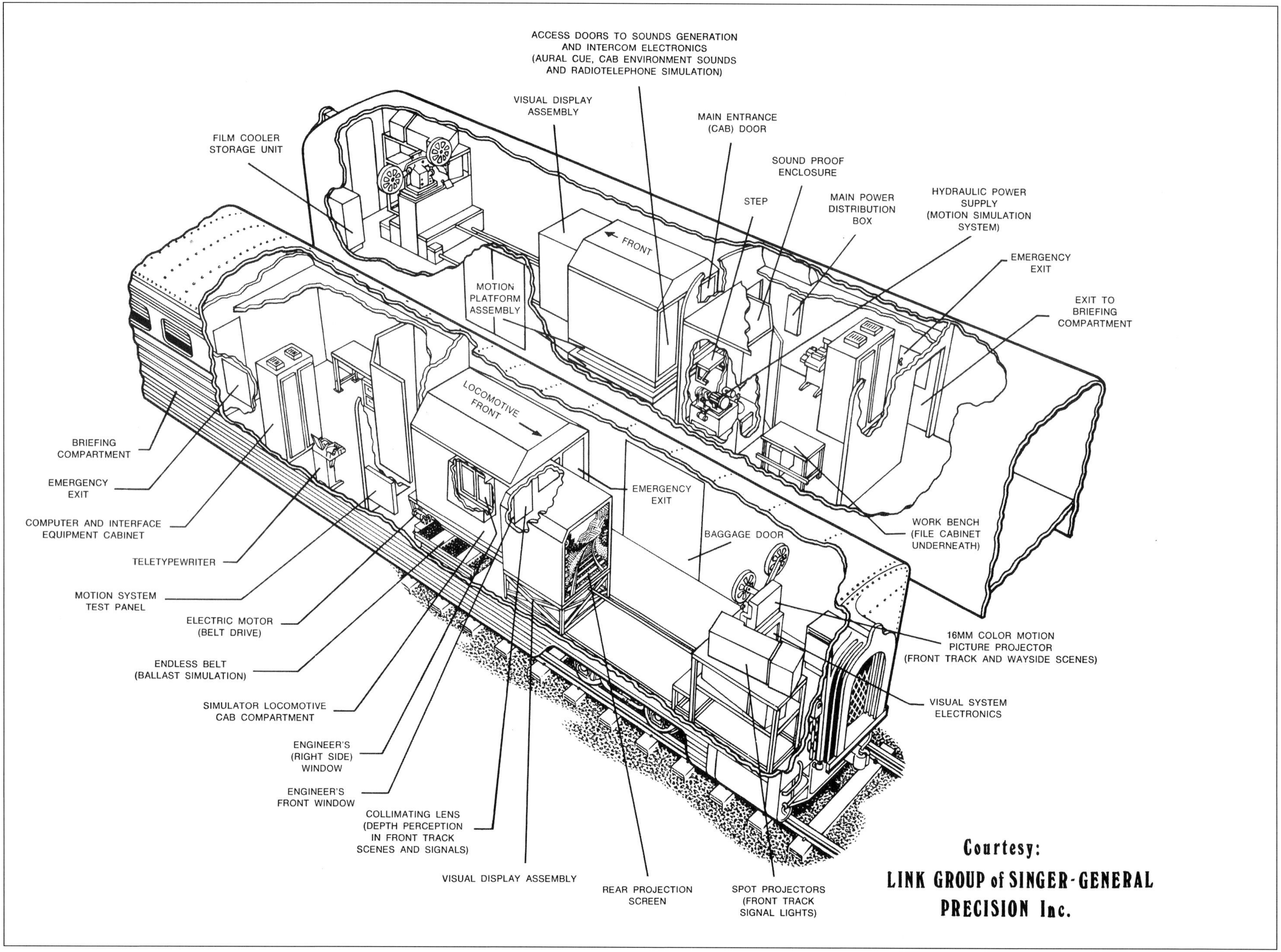
ACCESS DOORS TO SOUNDS GENERATION AND INTERCOM ELECTRONICS (AURAL CUE, CAB ENVIRONMENT SOUNDS AND RADIOTELEPHONE SIMULATION)
VISUAL DISPLAY ASSEMBLY
MAIN ENTRANCE (CAB) DOOR
FILM COOLER STORAGE UNIT
SOUND PROOF ENCLOSURE
STEP
MAIN POWER DISTRIBUTION BOX
HYDRAULIC POWER SUPPLY (MOTION SIMULATION SYSTEM)
FRONT
EMERGENCY EXIT
MOTION PLATFORM ASSEMBLY
EXIT TO BRIEFING COMPARTMENT
LOCOMOTIVE FRONT
BRIEFING COMPARTMENT
EMERGENCY EXIT
EMERGENCY EXIT
COMPUTER AND INTERFACE EQUIPMENT CABINET
WORK BENCH (FILE CABINET UNDERNEATH)
BAGGAGE DOOR
TELETYPEWRITER
MOTION SYSTEM TEST PANEL
ELECTRIC MOTOR (BELT DRIVE)
16MM COLOR MOTION PICTURE PROJECTOR (FRONT TRACK AND WAYSIDE SCENES)
ENDLESS BELT (BALLAST SIMULATION)
VISUAL SYSTEM ELECTRONICS
SIMULATOR LOCOMOTIVE CAB COMPARTMENT
ENGINEER'S (RIGHT SIDE) WINDOW
ENGINEER'S FRONT WINDOW
COLLIMATING LENS (DEPTH PERCEPTION IN FRONT TRACK SCENES AND SIGNALS)
VISUAL DISPLAY ASSEMBLY
REAR PROJECTION SCREEN
SPOT PROJECTORS (FRONT TRACK SIGNAL LIGHTS)
Courtesy:
LINK GROUP of SINGER-GENERAL PRECISION Inc.

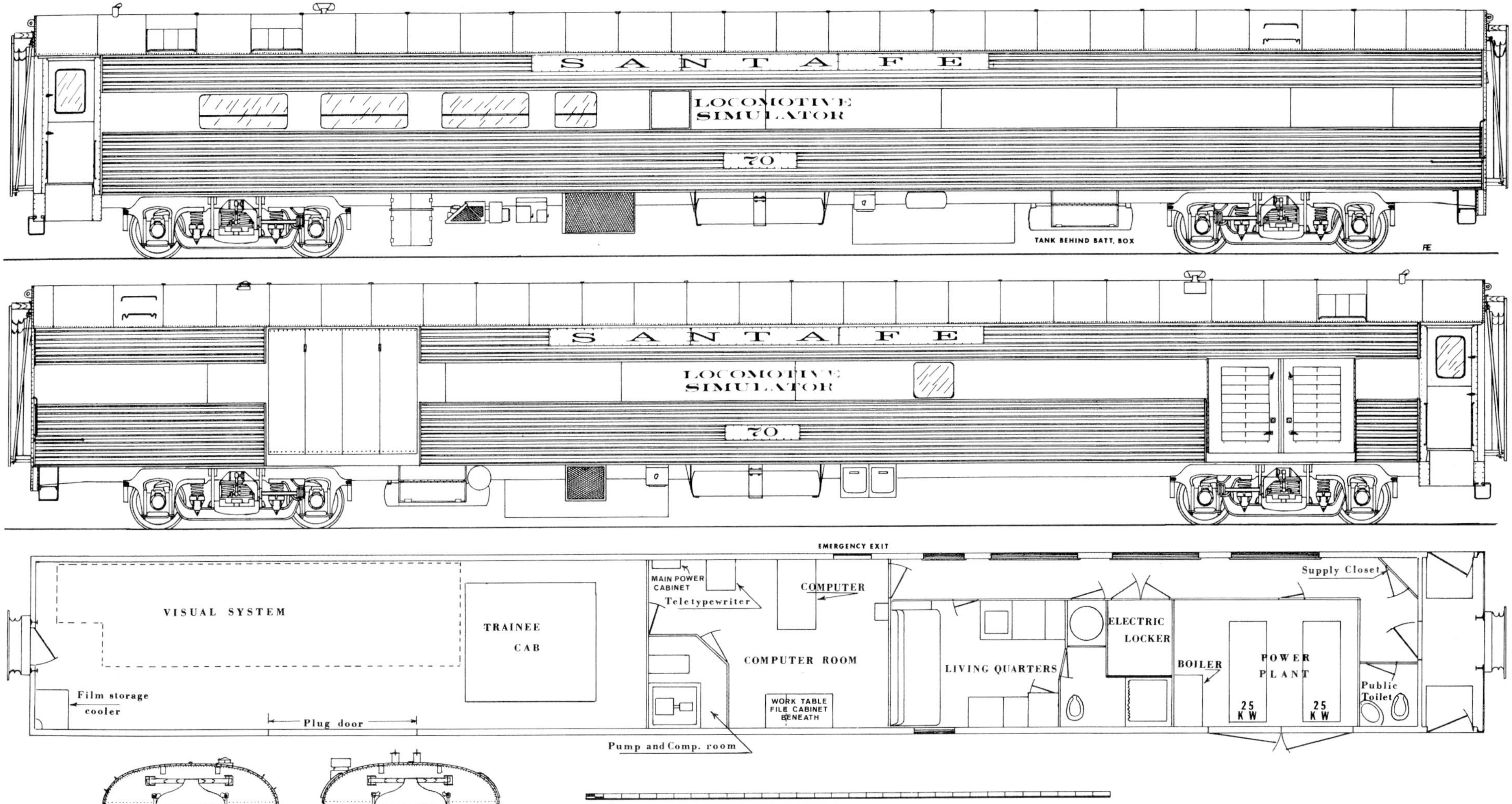

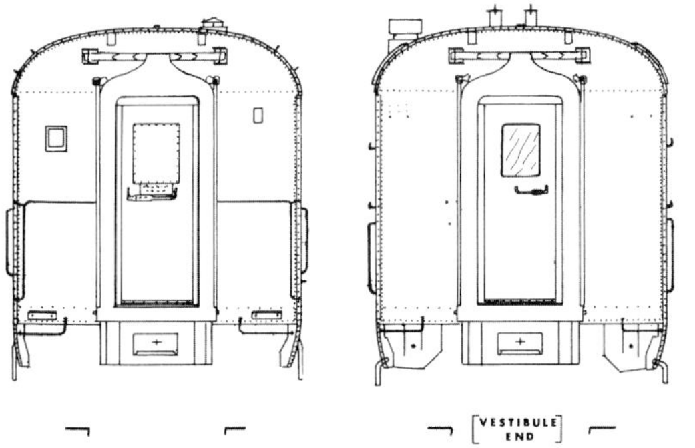

Locomotive Simulator Car 70

Pullman 1947

Rebuilt from Chair-Lounge 3179, 1969

Drawn by Frank Ellington, November 1974

5009 A.T.&S.F. Film Car

Film car 30 sits on the Business car track at Topeka, lower right, March 8, 1969, just after the initial modifications had been completed. One-half of the original open platform is still intact. In the two views above, the Film car, now numbered 5009, has the final, fully enclosed, platform at Topeka in July 1970. *-Two photos above, Frank Ellington; lower right, Joe McMillan*

Santa Fe's Film car, was converted from heavyweight Business car 30 at the Topeka Shops in March 1969 in anticipation of the new Diesel Simulator car, which would be released from Topeka a short time later. The primary purpose of the Film car was to allow cameras in a special housing to photograph the right-of-way under operating conditions as seen from the cab of a diesel. This film was then used in car 5008, the Diesel Simulator, to instruct prospective locomotive engineers.

Many test runs were made and a number of changes were made during the first few months of operation. During the initial runs, the Film car bore the original Business car number, 30. Coach green prevailed as the exterior color after the conversion. The old observation end was modified during mid-March 1969 by completely enclosing it, and the 30 emerged from the shops in July 1970 with light orange safety stripes on what was now the front end and a new number, 5009, centered on each side of the still coach green car. The next major change to the exterior transpired in 1972 when the complete body was repainted gray with silver roof and silver safety stripes. The 5009 was subsequently renumbered 73, Film car, in January 1973 at Topeka.

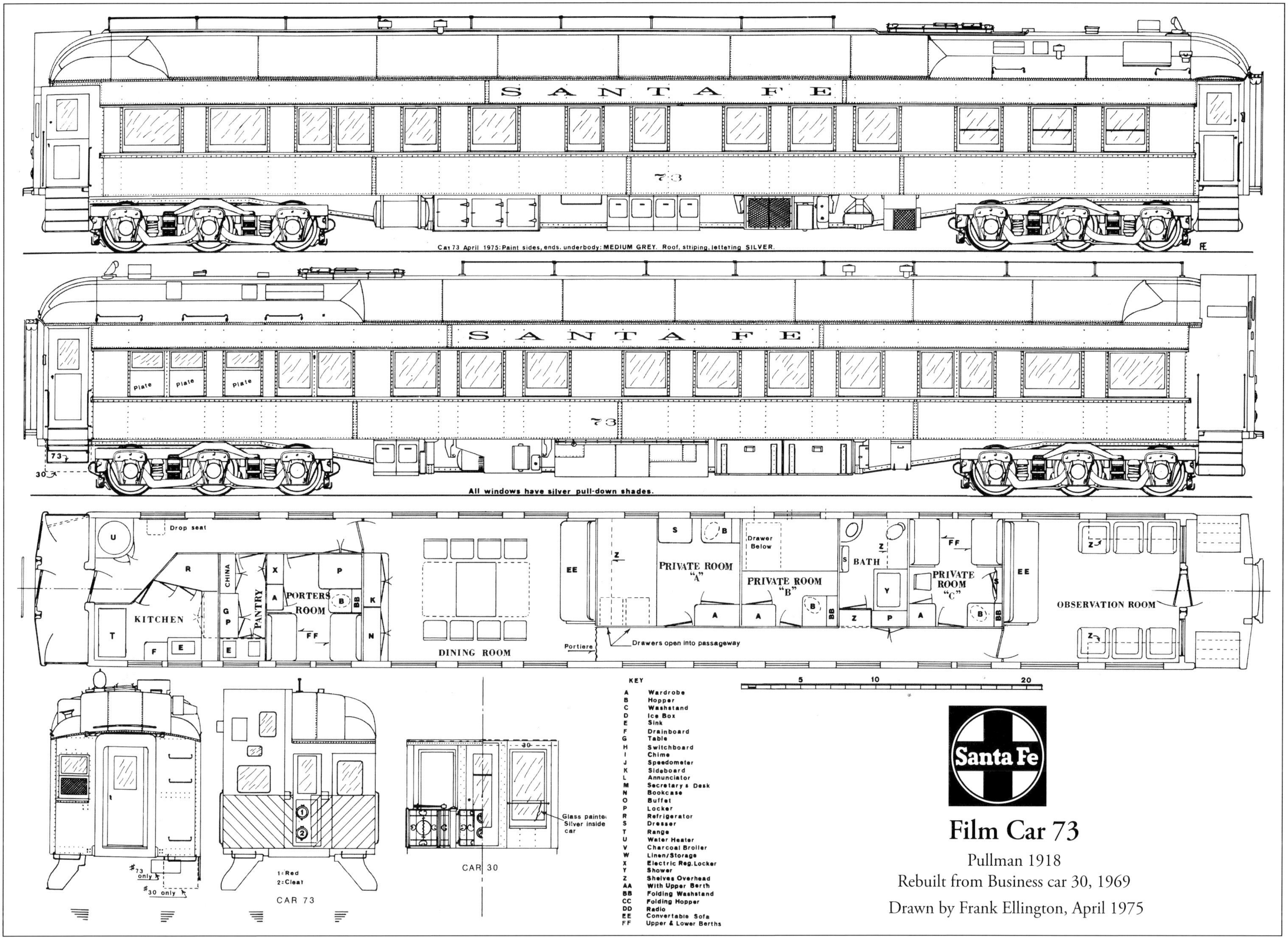

KEY

A	Wardrobe
B	Hopper
C	Washstand
D	Ice Box
E	Sink
F	Drainboard
G	Table
H	Switchboard
I	Chime
J	Speedometer
K	Sideboard
L	Annunciator
M	Secretary's Desk
N	Bookcase
O	Buffet
P	Locker
R	Refrigerator
S	Dresser
T	Range
U	Water Heater
V	Charcoal Broiler
W	Linen/Storage
X	Electric Reg. Locker
Y	Shower
Z	Shelves Overhead
AA	With Upper Berth
BB	Folding Washstand
CC	Folding Hopper
DD	Radio
EE	Convertable Sofa
FF	Upper & Lower Berths

Film Car 73

Pullman 1918
Rebuilt from Business car 30, 1969
Drawn by Frank Ellington, April 1975

The initial trips recorded with the Film car 5009 were six locations along varying portions of the Santa Fe system, including Taiban to Clovis, in eastern New Mexico; Fox to Domingo, New Mexico; LaPlata to Marceline, Missouri; Flagstaff to Williams, Arizona, and yard switching in Temple, Texas, and the 25-mile run out of Williams, Arizona, on the Grand Canyon line, a notoriously difficult stretch of track. (During the filming of the Grand Canyon line, four deer crossed the track and they were now a permanent part of the simulator training. It was not unusual for trainees to react to this scene with a couple warning blasts from the simulator's air horn.) The Film car was renumbered 73 in January 1973 and is seen here at Topeka in October the following year in a fresh coat of solid passenger grey, with silver safety stripes and silver roof. Having completed its purpose, it doubles as a Film-Classrom car and was eventually retired and sold to Kasten Railcar on February 8, 1988. It passed through several hands and as of this writing sits derelict in Cherryvale, Kansas. *-John B. McCall*

5010 A.T.&S.F. Air Brake Instruction

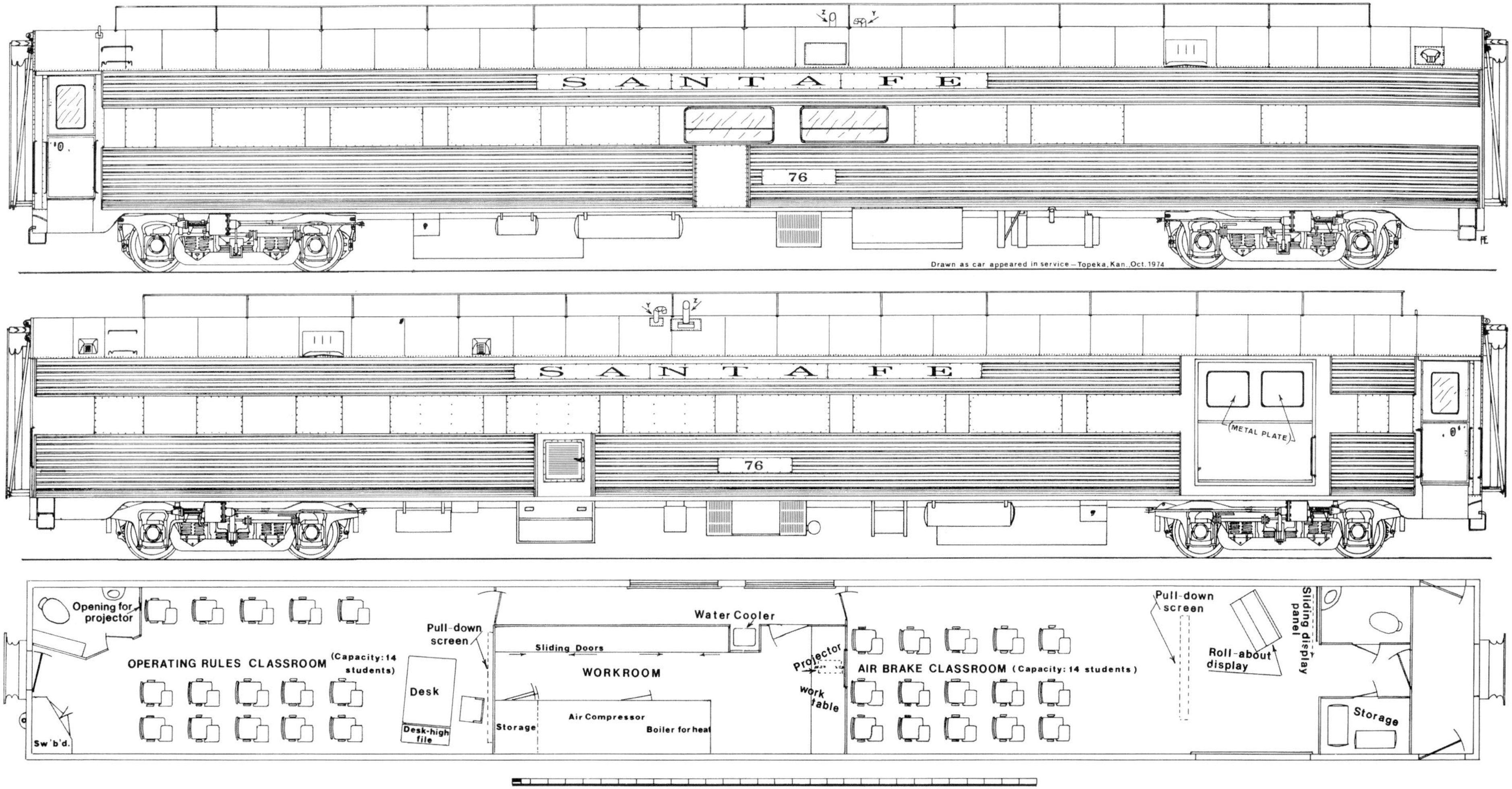

Air Brake Instruction car 5010 was converted from lightweight Vendo-Lounge 1346, originally Club-Lounge ATSF 1346, at Topeka in August 1972. Built by Pullman in 1950, the 85-foot car was renumbered 76, Instruction car, January 1, 1974, at Topeka.

Air Brake Instruction Car 76

Pullman 1950

Rebuilt from Club-Lounge 1346, 1972

Drawn by Frank Ellington, November 1974

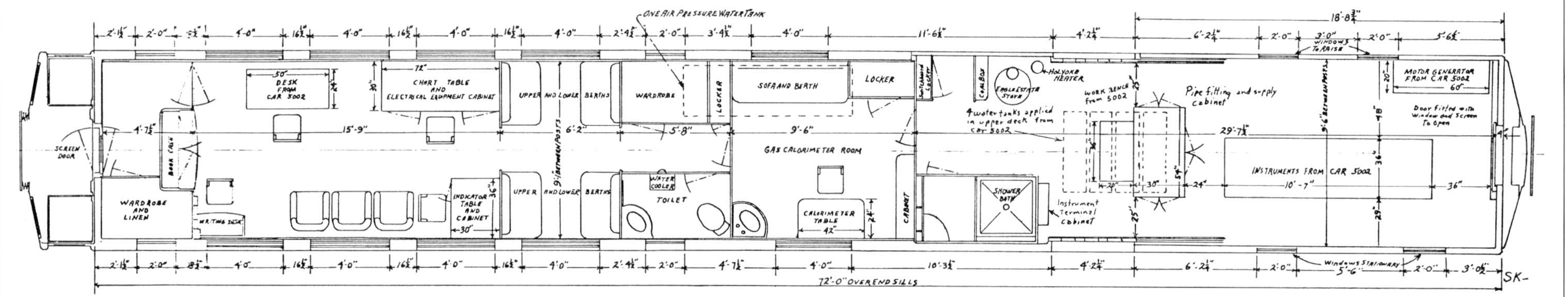

Power Plant Test Car 5015
(Rebuilt from Baggage-Buffet-Library 1338)
February 27, 1939

Dynamometer, or Test Car 5015 had its origins in this car, above, Power Plant Test car ATSF 5001, which was itself converted from 72-foot 1914 Pullman-built Baggage-Buffet-Library car 1338, named San Juan. The conversion to the 5001 took place at Topeka in 1939, where this photo was taken September 27th of that year. The 5001 was in turn fitted out as a Dynamometer car at Topeka in September 1961 to become ATSF 5015, Test Car. Aside from a few blanked out windows and MU electrical hook ups, the biggest change was the addition of bay windows on each side of the car toward the front. *-Two photos, John B. McCall collection*

Painted gray with a black roof, three year old ATSF Test Car 5015 is seen west bound at Summit, California, on March 4, 1964, cut in behind the road power on an unusually long 140-car Kaiser coal train destined to the steel mill at Fontana, California. *-Chard Walker*

Test Car 5015 was renumbered 83 on July 3, 1973. This is the opposite side of the car, seen at Topeka the same year. *-Joe McMillan*

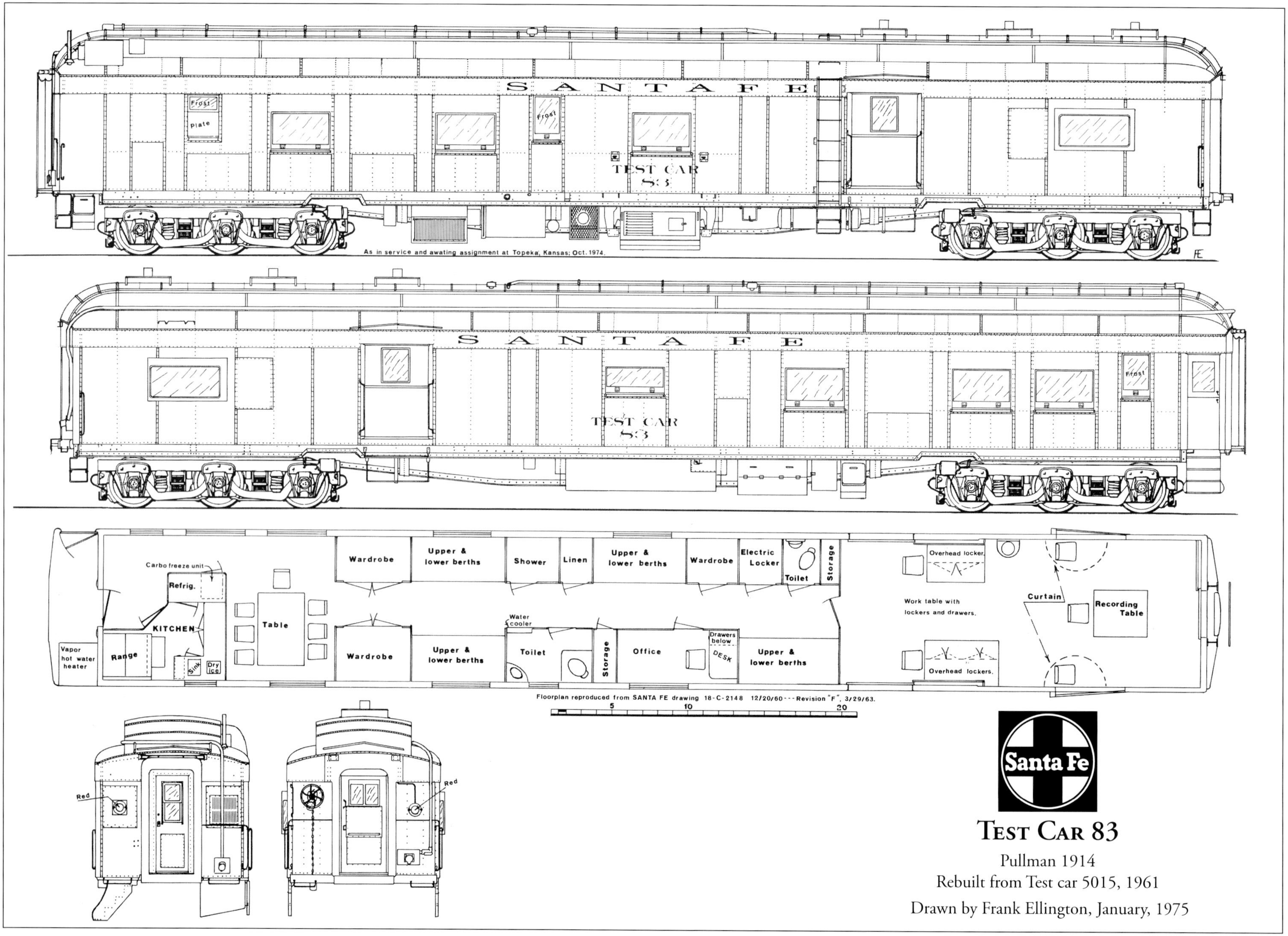

Test Car 83

Pullman 1914

Rebuilt from Test car 5015, 1961

Drawn by Frank Ellington, January, 1975

Although called a Test Car, the 5015 was in fact a Dynamometer car. While a spotter notes landmarks from the bay window of the 5015, a technician monitors the readout feeding from the dynamometer in a continuous stream. At right, the 5015 is seen cut in behind new SD45s to record train resistance measurements, truck and stress test. *-All, Santa Fe Collection, Kansas State Historical Society*

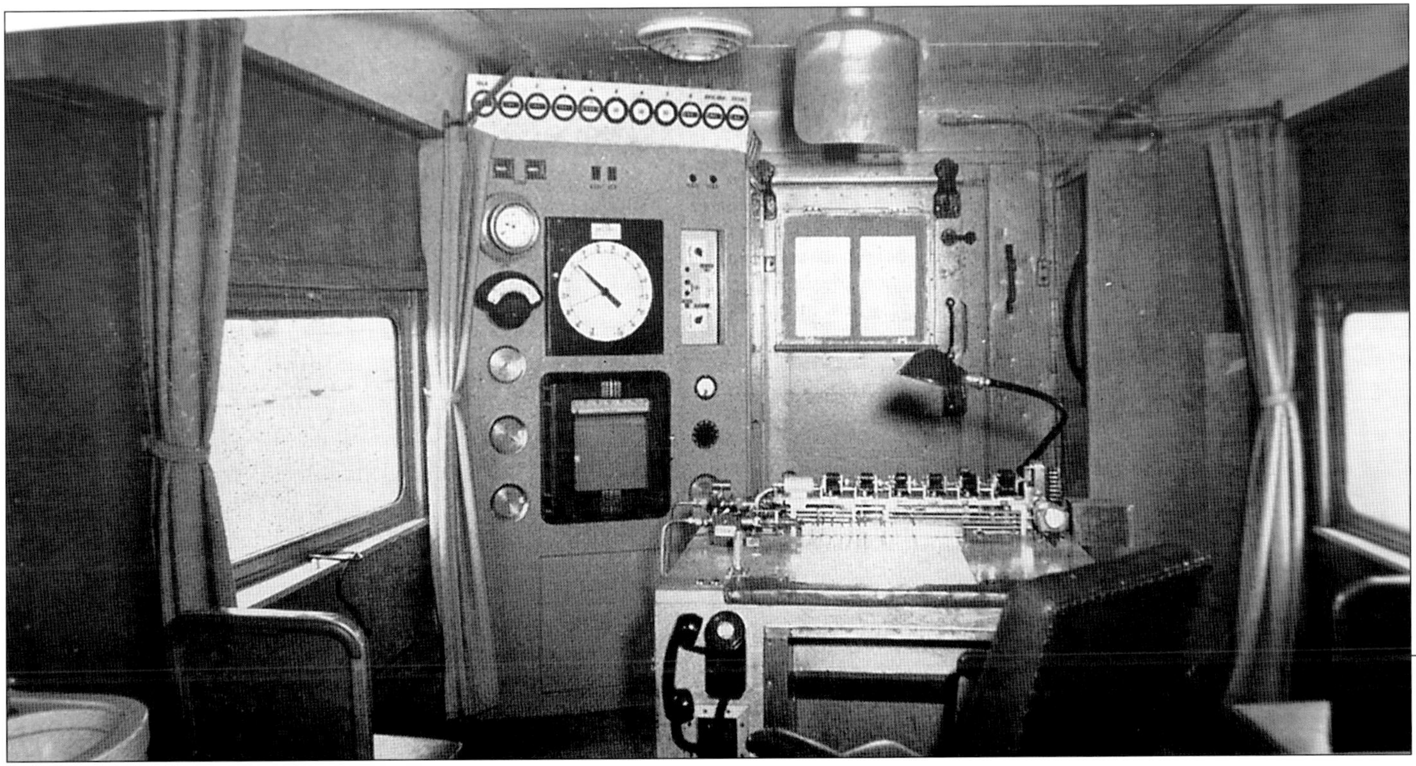

During March 1977 Test car 83 underwent some modifications and retrofitting at Topeka and emerged in solid silver paint as Research And Test Car 83. In addition to new signage on the sides, the car now had new bay windows, new sealed window units, new end windows and end doors on the instrument end, relocated electrical junction boxes and new red warning lights. The car is seen at Topeka, above, in June 1977, and six years later, below, on August 5, 1986. Note four-wheel trucks. The car remains in service as of this writing as BNSF 83, Research And Test Car, and is named Canyon Diablo. *-Left, Santa Fe photo, Railroad & Heritage Museum Collection, Temple, Texas; others, Bill Childers*

85 A.T.&S.F. Track Geometry Car

Santa Fe's Track Geometry Scientific Testing Car 85, placed in service in 1973, was at that time the most sophisticated of its type. Rebuilt from heavyweight Business car 39 in October 1973, it was scheduled to be numbered 5020, but while in Topeka Shops a renumbering program was instituted and it emerged as Track Geometry car 85. The 73-foot car, seen here at Topeka July 9, 1975, was modified to be self-sustaining, so it could operate in either freight or passenger train service. As of this writing, the car is serving BNSF Railway as BNSF 85, Track Geometry. *-Jim Primm photo, Joe McMillan collection*

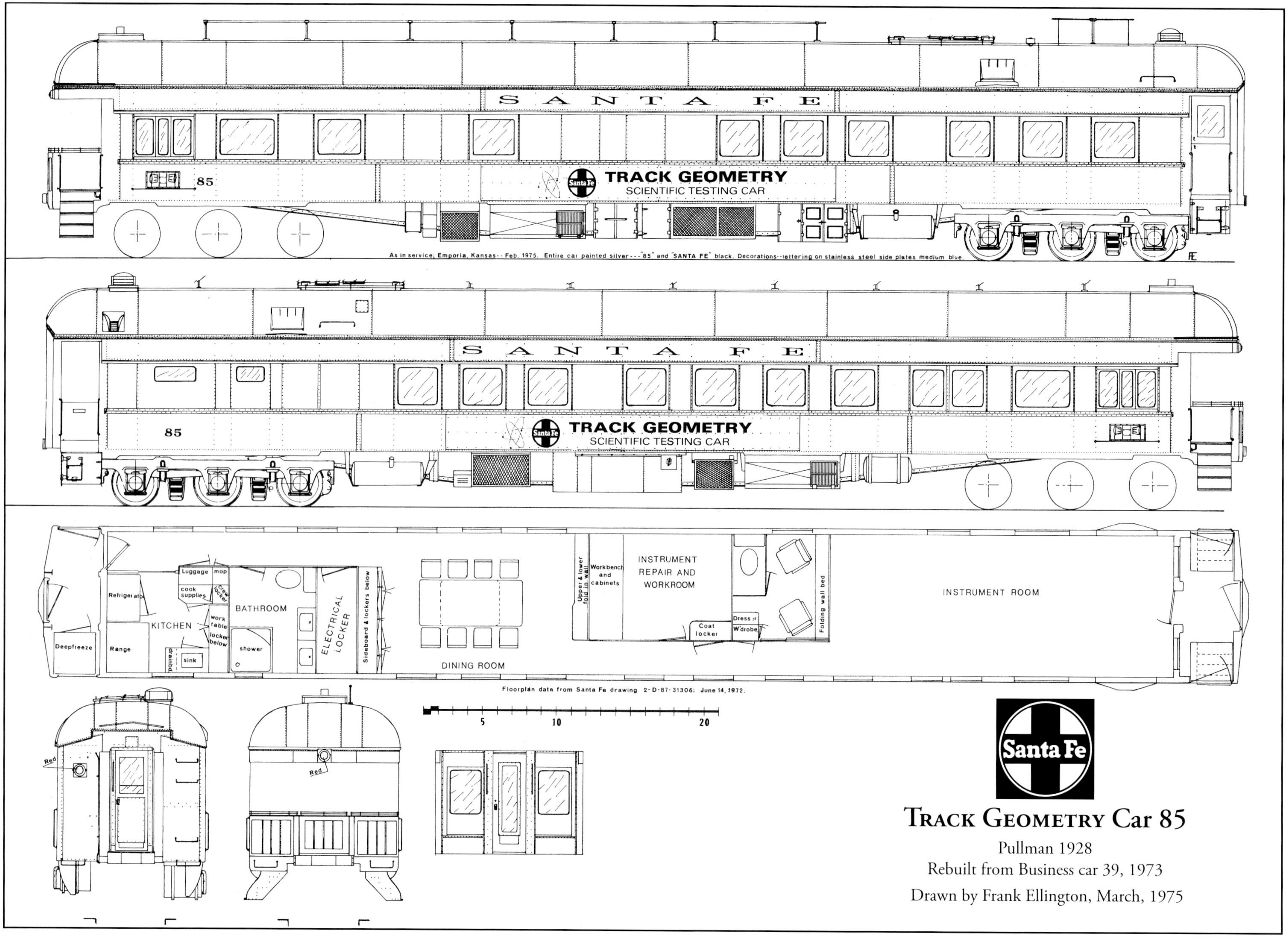

TRACK GEOMETRY CAR 85

Pullman 1928

Rebuilt from Business car 39, 1973

Drawn by Frank Ellington, March, 1975

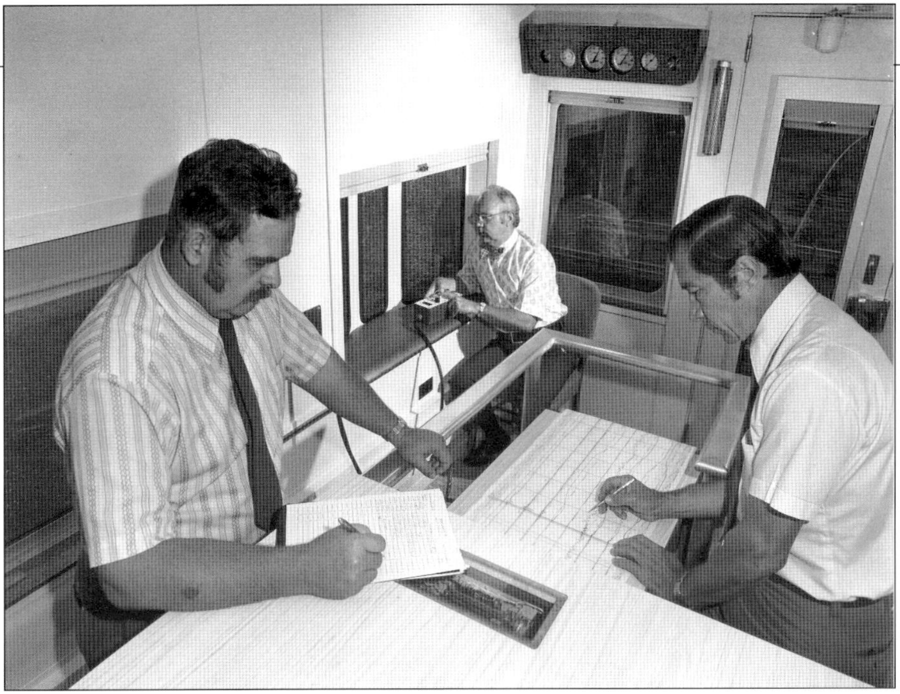

The business end of car 85 and a detailed view of a rear truck fitted with sophisticated equipment. Utilizing highly-sensitive measuring devices such as accelerometers, optical encoders coupled with contact feelers and a gyroscopic stabilized pendulum, the car constantly could monitor and record ten different track characteristics coupled with location identification by mile posts and other wayside landmarks. The measuring equipment was designed to be capable of day-to-day operation at speeds up to 80 m.p.h. with a minimum of down-time for maintenance and recalibration. Featuring an on-board mini-computer and the latest electronic sensing and recording devices, the car was the result of a joint design effort between Santa Fe and Rail Tech-Modern Machine and Tool of Newport News, VA. Track characteristics measured included surface and alignment of both rails, twist, gauge, superelevation, along with lateral and vertical acceleration, key benchmarks, distance and speed. An electrostatic recorder constantly displayed the measurements on an analog strip chart of the track structure. All deviations above selected minimum values were recorded digitally by the computer as to magnitude and location, which was used in studies to forecast and plan track maintenance and funding requirements. It also established a system master track data bank enabling management to correlate the rate of wear and traffic load factors and allowed computer simulation of various types of cars over the railroad. The front view of the car was taken at Topeka July 5, 1975. *-Above, Frank Ellington; above right, Santa Fe; right, Jim Primm photo, Joe McMillan collection*

By the early 1990s, TRACK GEOMETRY car 85 had received a new paint job of silver-gray with dark blue the length of the car's side window panels, as well as major changes to the electronic apparatus on the rear trucks. The car is seen at Bragdon, Colorado, above, on the Joint Line, on June 18, 1992, in company with GEOMETRY SUPPORT car 86. The latter car is seen, right, at La Junta, June 27, 1992. There is speculation that Car 86 was former Pullman/Topeka Shops-built semi-lightweight Baggage & Express car 3908, built in 1963 and its ultimate disposition otherwise unaccounted for. *-Above, Steve Patterson; right, Lee Burglund, SFRH&MS Collection*

Steam Generator Cars

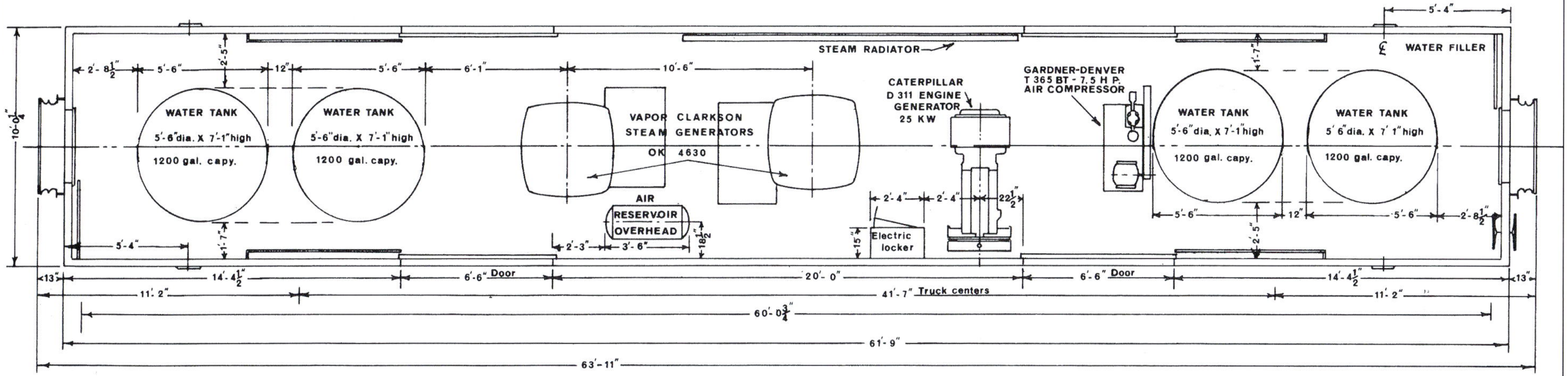

Steam Generator Car 9005
(Rebuilt from Baggage & Express 3707)
April 12, 1962

The first Steam Generator car, 9005, seen at Topeka, March 20, 1965, was converted from baggage car 3707 by the Topeka car shops in 1962. Unlike the steam generating tenders, 9000-9004, which required an extra fireman, the 9005 operated unmanned. It was equipped with two Vapor-Clarkson OK 4630 steam generators, a Caterpillar D311/25kw diesel engine and generator set for boiler pumps, and a Gardner-Denver air compressor for fuel atomization. The 9005 has a water capacity of 4,500 gal. and a fuel capacity of 800 gal. It was renumbered 6 in 1968 and finally 130 in 1970. This was the only such car to be shadow-lined.
-Joe McMillan

Another view of the first car, 9005, a couple years later, at Ft. Worth in July 1967. This is an interesting view as the car's doors are open giving a glimpse of the interior. Note that the car has by now lost its shadow-lining. *-John B. McCall*

There were ten more Steam Generator cars constructed by the Topeka Shops between March and July 1970, rebuilt from the longer 3900-series Baggage cars. Their internal components were identical. These Steam Generator cars were used extensively after mid-1970, and into the Amtrak era on the combined *Super Chief/El Capitan* and the *San Francisco Chief*. Amtrak leased several of the Steam Generator cars and for a short period to be used with their *Super Chief/El Capitan*, prior to the delivery of Amtrak's SDP-40Fs. At times one of the steam generator cars was operated at the rear of the consist for added steam for heating. The Steam Generator cars were also used in conjunction with Business cars behind freight trains, such as this setup below, with Steam Generato car 131 sandwiched by Business cars 31 and 38 behind the *Super C* at Needles, California, February 13, 1972. They were also used as stand-by boilers for terminals, when a steam boiler plant was temporarily out of service. By 1985, Santa Fe had discarded all but two of their Steam Generator cars (138 and 139), which were kept on hand to provide steam for the company business train. The Business cars were gradually converted to on-board power for heating, lighting, and air conditioning and the two cars were no longer necessary. *-Joe McMillan*

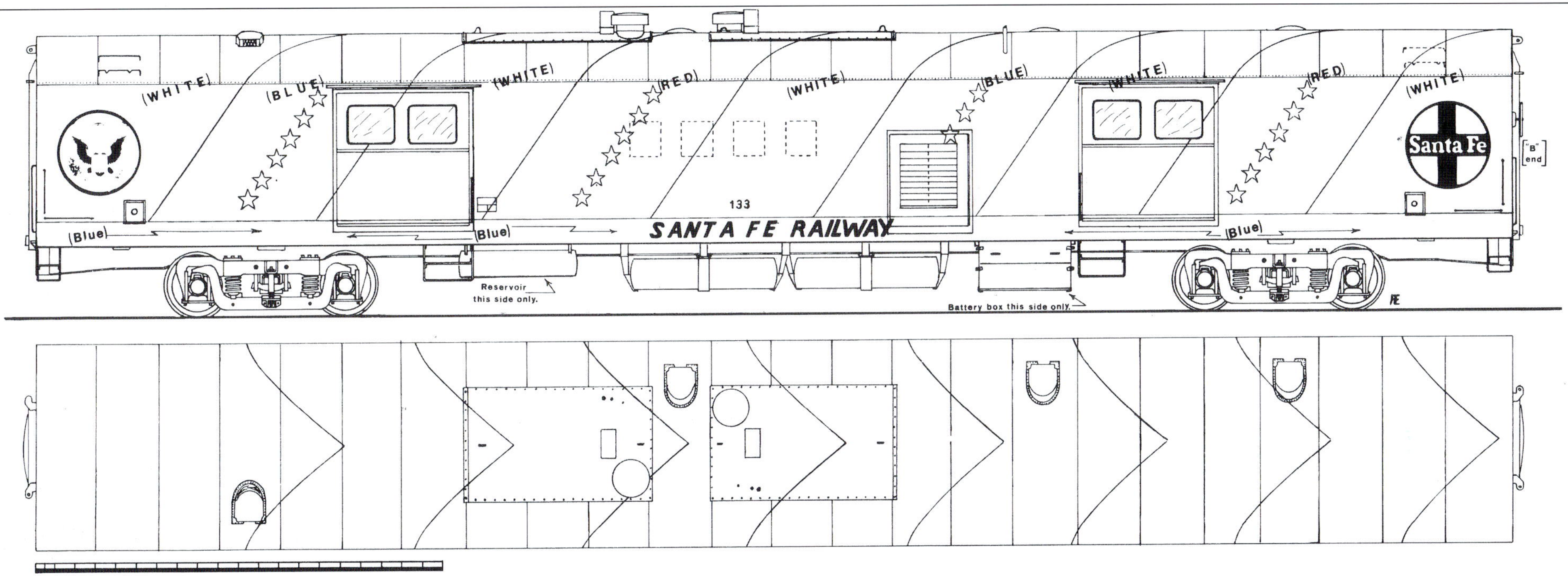

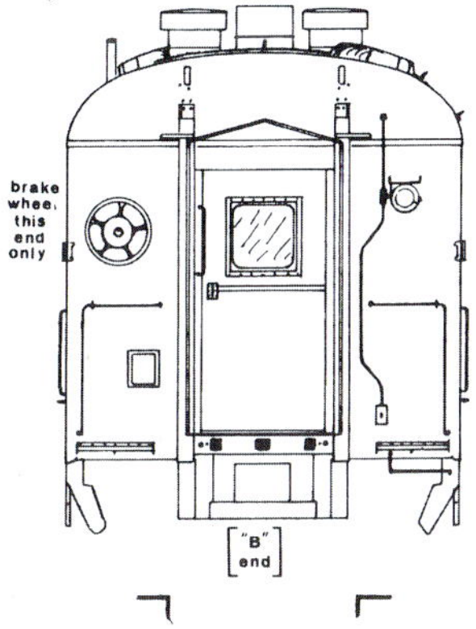

Steam Generator Car 133

Pullman/Topeka 1965
Rebuilt from Baggage & Express car 3922, May, 1970
Drawn by Frank Ellington, March, 1974

Steam Generator car 133 lays over at Barstow, California, in January 1973. Steam Generator 133 was painted red, white and blue for the "Spirit of 1776" train in late-1972 and was still operating with this paint scheme as of June 30, 1975. *-Seeth "Corky" Thompson*

Steam Generator car 132, built from Baggage car 3920, was equipped with steam-jet nozzles for blowing sand from the tracks in the California desert in 1971 and is seen at Needles that August between runs. This equipment was later transferred to Steam Generator car 134. *-Joe McMillan*

With water a premium in the California desert, provisions were usually made for an adequate supply. Here Steam Generator cars 134 and 130, an auxiliary Water car, RSD15 9844 and a Caboose comprise a work train engaged in blowing sand from the tracks, one mile west of Cadiz, California, on January 20, 1973. *-Joe McMillan*

Steam Generator Tenders

Between March and December 1943, Santa Fe converted five 1337 Class steam locomotive tenders into Steam Generator cars to be used to heat and air-condition passenger trains powered by diesel freight locomotives. The cars, numbered 9000-9004, were manned buy an extra fireman and were coupled to the rear of motive power consists.

The cars contained two model CFK 4225 Vapor-Clarkson steam generators. Two dynamos (turbine-generator sets) for supplying electrical power needed to run the steam generators were located just outside the steam generator room on top of the water tank. The turbines were piped into the steam lines from both steam generators so that either turbine could be operated from steam produced by either steam generator. It required all the steam from one steam generator to operate the turbine, then it ran at low speed.

The fuel tank was mounted on the underframe of the unit and was filled from either side through a filler pipe. An emergency fuel cut-off valve was located beneath the tank and could be tripped by using anyone of three cables located on each side of the unit and one in the unit near the fuel pump. The water storage tank was filled from a manhole in the top of the tank.

Outside of the front of the unit to the right of the doorway was a receptacle to accommodate a jumper cable so that batteries of the locomotive unit ahead could be trainlined to the tender unit to operate one steam generator (two steam generators, if two locomotive units were jumpered together).

Air for the air reservoir was supplied from the main reservoir on the diesel locomotive. This reservoir was located at the right front corner of the steam tender and supplied air for atomization of fuel going into the steam generators.

A fuel measuring gauge marked in gallons was located near the rear wall of the steam generator room at floor level. The water tank had no gauge and was checked visually by looking into the water tank.

Fresh out of the shop in March 1943, Steam Generator car 9000 has just been converted from the tender off 4-6-2 1367. Built to provide heat and air conditioning to passenger trains handled by freight diesels during the busy war years, 9000 is painted in freight diesel blue and yellow with the three red stripes, just like the early version of the freight FT units. The car was renumbered 1 in August 1967 and retired June 13, 1968. *-John B. McCall collection*

The steam tender was equipped with the standard alarm signals so that in case of steam generator failure the alarm bell rang and a green signal would light in the steam tender and each unit of the diesel locomotive. On locomotives having two lead units (double end operation), such as A-B-B-A, the alarm circuit was not established rendering the alarm light and bell inoperative, due to no control jumper cable. It was possible to determine a defective condition by visual observation of the safety devices which automatically shut the steam generator down.

Steam Generator car ATSF 9004, built from the tender off 4-6-2 1357 in December 1943, trails freight diesel 161 on a troop train at San Bernardino, California, on February 10, 1946. The 9004 was renumbered 5 in August 1967 and retired June 13, 1968. *-Jack Whitmeyer, SFRH&MS Collection*

Steam Generator 9001, formerly the tender off Pacific 1360, is seen at San Diego in 1954 providing steam for the Hi-Level display train. Note that it has lost its red striping. The 9001 was renumbered 2 in August 1967 and retired September 27th the same year. *-Seeth Corky Thompson*

Steam Generator 9000, formerly the tender off Pacific 1367, is seen at Ft. Worth on November 10, 1963, in what appears to be stationary boiler service. The car has lost its red striping and will be renumbered No. 1 in August 1967 and scrapped the following year. *-Dick Kuelbs*

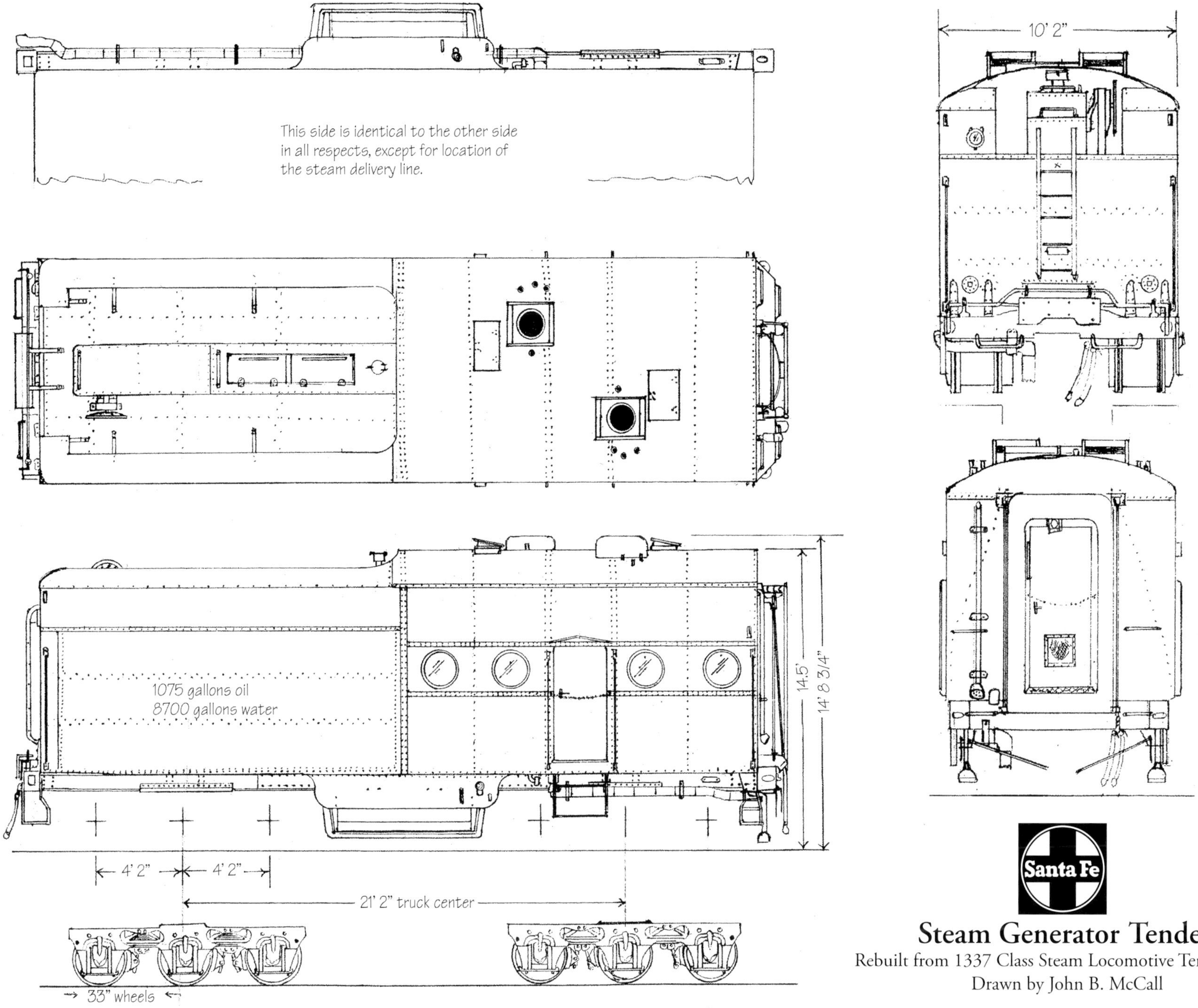

Steam Generator Tender

Rebuilt from 1337 Class Steam Locomotive Tender, 1943

Drawn by John B. McCall